How Deep Is the Lake

Caitlin Press Inc.
8100 Alderwood Road,
Halfmoon Bay, BC V0N 1Y1
www.caitlin-press.com

Cover design by Vici Johnstone
Typeset by Demian Pettman
Printed in Canada

Caitlin Press Inc. acknowledges the Government of Canada, the Canada
Council for the Arts, and the British Columbia Arts Council for their
financial support for our publishing program.

Library and Archives Canada Cataloguing in Publication

O'Callaghan, Shelley, 1948- author
     How deep is the lake : a century at Chilliwack Lake
/ Shelley O'Callaghan.

ISBN 978-1-987915-39-6 (softcover)

     1. Chilliwack Lake (B.C.)—History. I. Title.

FC3845.C455O33 2017   971.1'37   C2016-907717-9

How Deep

is the Lake

A Century at
Chilliwack Lake

Shelley
O'Callaghan

CAITLIN PRESS

To my grandchildren. I entrust the future of the lake to you.

Previous page: View from the author's cabin down Chilliwack Lake with the snow-capped mountains at the south end in the United States. This view has not changed over the family's one hundred years at Chilliwack Lake. Courtesy of O'Callaghan family.

CONTENTS

MAP

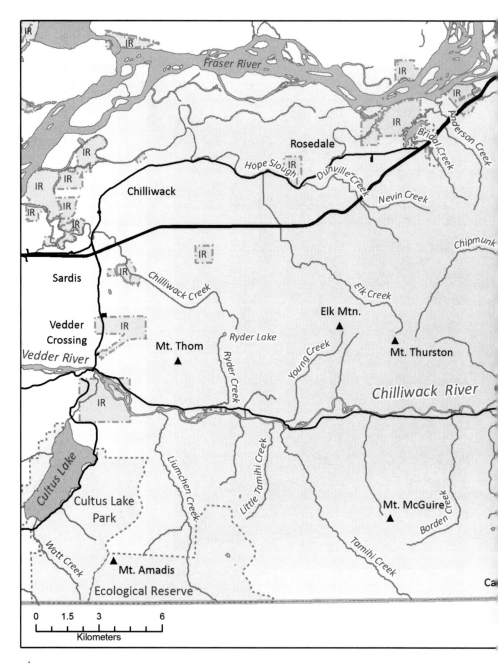

★ on the inset map shows location of Chilliwack Lake (now a provincial park) in British Columbia, Canada. Area map provided by GIS technician Shannon Sigurdson of the Fraser Valley Regional District.

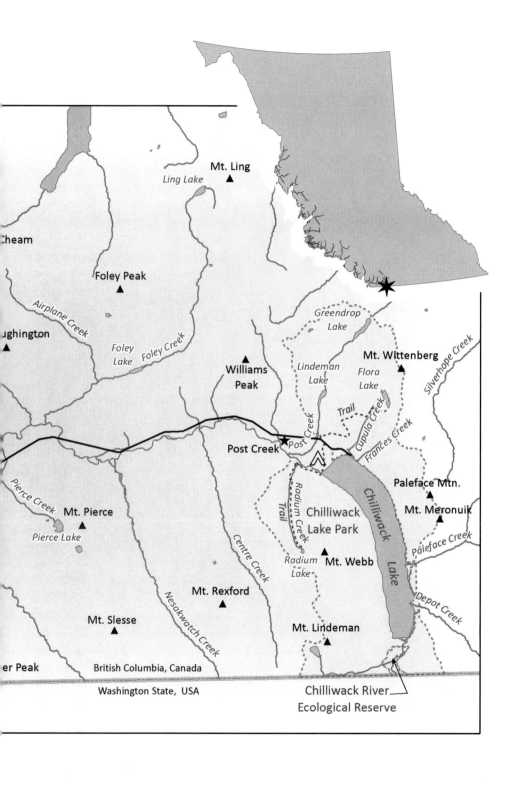

Author's Note

As much as possible, I have presented events, people and places as they exist or existed. Timing of some events was changed to assist the narrative. When I was not able to ascertain facts, I created the particulars of individuals' lives as I imagined they would have experienced them. The stories of the early life of my family are presented to the best of my memory. The memories of others will likely differ from my own.

I was particularly concerned with using current terminology when talking about the first people of Chilliwack Lake. The reference texts used various names, depending on the era in which they were written. Through my recent research, I understand that the first people of Canada wish to be called Indigenous peoples. I wish to be respectful and so have used that term.

Canada converted from the imperial measurement system to the metric system in 1970. I use the metric system, but retain the imperial system in those historical instances where imperial measurements were used.

I relied on a number of sources made available through the Chilliwack Museum and Archives, the Stó:lō Research and Management Centre and the Royal BC Museum and Archives. Two texts that I found particularly useful were: *In the Arms of the Mountains: A History of the Chilliwack River Valley,* published by the Chilliwack River Valley Historical Society, and *A Stó:lō Coast Salish Historical Atlas,* edited by Keith Thor Carlson.

This story cannot exist in words only. Without pictures, how could anyone understand the beauty of this place that grabs you by the heart. Many of the photographs are taken from my own photo albums and those of our extended family. Others whose family came early to the lake—the MacLeods, the Coulters, and Kris Sanders—generously provided their visual memories. Other photographs have been made available by the Chilliwack Museum and Archives, the Royal BC Museum and Archives, and the Stó:lō Research and Management Centre.

Opening the Cabin

Opening day: April 12. I start my lists. The basics: sugar, flour, eggs, coffee and milk. Meals for the weekend: spaghetti, grilled cheese sandwiches and tomato soup, and pancakes. Bedding that we brought home in the fall to wash, then stored in plastic bins that sit on shelves in the garage with THE LAKE written neatly in bold letters on the sides. Just seeing the word fills me with anticipation.

The Lake.

For sixty-six years I've been coming to this sub-alpine teardrop of water in the Cascade Mountains of British Columbia, the third generation of my family to spend her summers here. My grandparents' trek in the late 1920s took two days. They stayed overnight in the town of Chilliwack, leaving by car at dawn for the logging railway that snaked its way 32 kilometres up the Chilliwack Valley. At the rail depot, boxes of powdered milk, tins of butter, sacks of flour, sugar and salt in rough cotton bags sewn up at the top, cabbages, onions, potatoes, and carrots in burlap sacks, waxed boxes of molasses and cans of kerosene were secured onto a speeder car, the motorized handcar that railway crews used to maintain the line and that carried people and supplies to the logging camp up the Chilliwack River.

Gran and Poppy stood on the back of the speeder car, behind their mound of supplies, and grasped the outer handles, the wind in their faces. The car followed the twists and turns of a narrow-gauge track that hugged the banks of the river to the rail's end, where a pack train of horses stood waiting for the boxes and duffle bags to be cinched on their backs. In single file, the riders led the horses with their loads the final 6.5 kilometres up a switchback trail, arriving at the lake by late afternoon. Charlie Lindeman, a trapper who lived at the lake year round, would be waiting. Poppy retrieved his rowboat from Charlie's boathouse and together they loaded the supplies into the boat for the ten-minute row along the north shore of the lake to the stretch of white sand that fronted the clearing where my grandparents had built their one-room log cabin under the towering Douglas firs. They arrived at nightfall, 48 hours and what seemed like a million miles from the city of Vancouver.

The packhorses and speeder car were before my time. From the time I was six weeks old until I was twelve, I arrived with my parents by float plane in

The author's mother and grandfather (far right) travelling by packhorse to Chilliwack Lake, circa 1935. The trip took eight hours following the trail from the Bell's Farm in the Chilliwack Valley to the north end of the lake. Courtesy of Webb family.

mid-July and returned home to Vancouver a month later. After a road was pushed through, we drove in, the ten-hour trip gradually whittled down as the highway improved.

Now, we make the journey from our home in North Vancouver in just two hours.

Seated beside my husband Patrick, the Jeep piled with many of the same provisions that Gran and Poppy had, we leave the Pacific Coast and drive east on the Trans-Canada Highway along the valley of the Fraser River. The Cascade Mountains are a distant backdrop to the rolling farmlands of the valley. The snow-topped peak of Mount Baker is visible eighty kilometres to the south in the United States. We pass the towns of Langley, then Abbotsford, and take an exit south towards Vedder Crossing, a town of ten thousand where we come in the summer for supplies and Blizzards at the Dairy Queen.

My shoulders loosen as we exchange the traffic of the highway for the quiet country road. In twenty minutes, just before entering Vedder Crossing, we take the bridge over the Chilliwack River and turn east again onto the Chilliwack Lake Road. Only forty kilometres to go. The road is paved but narrow, and hugs the Chilliwack River winding through the Chilliwack River valley. Between Vedder Crossing and the lake, we climb 600 metres in elevation. The river is deep green, turbulent with swells and white caps from the heavy rains of the past month. Fields dotted with farmhouses give way to forests of Douglas fir. Following the river valleys, first the Fraser Valley, then the Chilliwack River valley, we wend our way into the mountains. As the civilized world recedes, the peaks of the Cascade Mountains, once faint on the horizon, loom large in front of us. The world in which I feel most at ease takes over: river, forest and mountains. We press higher into the mountains and I give Patrick's arm a squeeze. We both smile. Soon we'll be at Chilliwack Lake. My favourite place on earth.

Patrick fell in love with the lake, too, on his first visit nearly fifty years ago. We vowed that one day we would build our own cabin down the shore from my parents and grandparents. In 1970, immediately after university, we married and travelled to Africa to spend two years as volunteer teachers with Canadian University Services Overseas in a rural boarding school in southern Zambia. We took almost nothing with us, but I packed a small, framed photo of the lake. I would pick it up and close my eyes, transporting myself halfway around the world to this piece of wilderness. Our second son, Michael, was born in Toronto and our first, Kevin, was born in Zambia. After finishing university he, too, spent a year teaching in southern Africa.

"Don't let anything change at the lake," Kevin said as he left for Namibia.

Kevin and Michael are now in their forties, each married with three children. They both live in North Vancouver, close to us. We are converging on Chilliwack Lake on this chilly April day to open the cabin, a family ritual that goes back to my grandparents' first days there. A tradition nearly a hundred years old.

The paved road ends at the entrance to the Chilliwack Lake Provincial Park. A narrow gravel forestry road continues straight ahead, then veers down the eastern shore to the south end of the lake. The border between Canada and the United States is three kilometres further on, accessible only by foot. Patrick slows the car to a crawl as he steers around and through the ruts and potholes of the forestry road. These few minutes seem to stretch forever. I can't wait to get there. I get my keys out of my purse, jangling them in anticipation. Finally, we turn south off the forestry road onto the access road to our three family cabins and to the cabin of our neighbours, the MacLeods. Patrick stops the car and I jump out to unlock the rusty single-bar metal gate, hanging on as the car pulls through, then pulling it closed. Now, I can truly shut out the world. I stand for a moment by the gate, breathing in the sharp scent of fir. Fingers of light poke through the forest canopy, illuminating patches of emerald moss on the forest floor. Douglas fir and western redcedar branches lean over the narrow dirt track. We drive through the bower of green halfway down the access road. If we kept going straight, we'd end up at the MacLeods. Instead, we turn east. Patrick stops the car again so I can open the second gate. Now we are on my family's land. The road goes straight for another kilometre to our family's cabins, with a lane to my parents' cabin, called Sandpiper. Another runs to my brother Chris's cabin, Cupola Lodge, an expansion of my grandparents' original log structure. Straight ahead is the lane to our cabin, Creekside.

The two-storey log house that greets us is not really a "cabin." It's more like a lodge. Patrick and I built it twenty-five years ago with dreams of the extended family that has, in fact, materialized. A wrap-around porch extends to form a prow-like deck overlooking the lake, perfect for happy-hour drinks and games of cards sitting on towels in the sunshine. The cavernous basement is big enough to store boats, paddleboards, life jackets, wetsuits, and bedding for the tent and boathouse. Our sons and their wives take two of the bedrooms upstairs. The grandkids cram into a large front bedroom lined with bunks. A small room just large enough for a single bed is reserved for Patrick's mum, called GG by the great-grandkids. She is often with us in the summer. Twelve years ago, we added a self-contained cabin we call the Brook, connected by a breezeway to the main cabin: an oasis of calm for Patrick and me. Two of the grandkids bed down in the boathouse in the summer. Overflow guests sleep in tents.

Officially, our land is Cupola Estates, the name of our family's 27.5 hectares, owned jointly by my brother Chris and me. The three "cabins" are built in adjacent clearings that front on a beach at the east side of the north end of the lake: we make a giant-sized footprint compared to the single one-room log cabin my grandparents built almost a century ago, one of only four structures on the lake for forty years.

Recently, I retired after twenty-five years as an environmental lawyer. My job involved helping resource companies, mostly in forestry and mining, to implement strategies for managing environmental risk. Researching the pieces of the social, geographic, economic and political land use puzzle was a key component of my work.

Now I want to turn my research skills to Chilliwack Lake, this place of tranquility and binding ties. Our family loves this retreat. We always have. It seems like part of our DNA. But we have never talked much about what the place means to us. I know almost nothing about its history. Why isn't that story part of our family folklore? Weren't we interested in the past? Or did we just not want to talk about it?

When Patrick and I first built Creekside, I looked for photos of early days at the lake. I found a few black-and-white photos and had them framed to hang on the walls. That's as far into the history as I have delved.

My six grandchildren have reached the age where they are asking questions. I have few answers. Although my family has lived here for nearly a century, there are no family diaries, few photos, and only my recollection of tales told to me by Gran and Poppy. My mother, who would have had stories to tell, is ninety-six now and lives in an extended care home in Vancouver. She can no longer remember either the past or the present.

Finally, Creekside comes into view. Patrick pulls the Jeep up close to the back porch so we can unload the car. I climb the back steps, unlock the cabin door and walk inside. A familiar smell greets me, a mixture of fir logs and crisp cold air. I open the sliding doors to the front deck and gaze at the blue-green oval of water nestled in the mountains. Its reflection shimmers through my soul. Chilliwack Lake. I am back.

I hear the beeping of horns and look out the back window to see two cars coming around the bend. A skiff of snow still whitens our access road and flattens the grass in patches behind the cabin. Patrick and I go out on the back porch to greet Kevin and Michael and their families. The grandchildren spill out of the cars and call excitedly to each other, and to us.

"Did you see the deer on the road at the first gate?" William asks. Will is eight, the youngest of Kevin's children. His nickname, used just within the family, is Scooch. His brother Finlay, called Fin, is eleven and his sister Bronwen, thirteen.

"Yes, and two fawns," I turn to Kevin. "How was your trip?"

"Okay. Traffic wasn't bad."

Kevin and his wife Joanna start to unpack the mini-van, corralling the three children into helping before they rush off. Bronwen sweeps her wavy brown hair over one shoulder and stoops to pick up a couple of bags of groceries. She shares my petite frame and determined personality. "Good things come in small packages," I often say to her.

"Should I put these away, Rani?" she asks, her voice rising above the hubbub.

"Sure. You know where they go."

I like the name they've given me, Rani. When our eldest grandchild, Nora, was two, she couldn't pronounce the G at the start of Granny so she just left it off. The name stuck.

Will is peppering me with questions.

"What are we going to do first? Is the water cold? What's for dinner?"

"Okay, too many questions," I say, giving him a hug. "Dinner's spaghetti and salad."

"Yay," he says.

Michael and his wife Molly are right behind Kevin's family. They have two girls and a boy: Nora, Fiona and Roger. Nora, the oldest of the grandchildren, is fourteen and has suddenly become a young woman. She bumps Will out of the way and gives me a hug. She stands close, sizing me up.

"I think I'm taller than you now, Rani," she says. "Let's measure!"

Since the kids were little, we've tracked their height on the bathroom wall. "After we get unpacked, okay?"

Nora turns to Patrick.

"Guess what Poppy, I passed my boat licence test!"

"Oh, oh," says Patrick. "Look out everyone. Stay clear of Nora driving the boat."

Nora rolls her eyes. "Poppy, you're the only one who doesn't have your licence."

I also like that they call Patrick Poppy. Kevin always called his father "Pops" and the toddler, Nora, chose "Poppy" for her grandfather. The name is nostalgic for me: I called my grandfather Poppy too.

Fiona, who is only a year younger than Nora, gets out of the car, stretching her long arms and legs like a cat. She shakes out her thick blonde hair and twists it into a knot on the top of her head. She's thirteen, the same age as her cousin, Bronwen. Roger is the same age as his cousin, Fin. Roger always makes me smile. He marches to the beat of his own drum. Today he's dressed in natty shorts, refusing to acknowledge the crisp April air.

"Has anyone been in the lake?" Roger asks, giving Patrick and me a big hug. Then he spies Fin and the cousins fall to the ground, wrestling.

As always, both my sons bring their dogs, Murphy and Dennis, beautiful retriever crosses called golden doodles. They run in circles nipping at each other's heels, happy to be released from the cars. I love these dogs. My "granddogs," as I call them. I look forward to having them stay with me in town and at the cabin whenever their families are away.

We all pitch in to unload the endless bags, boxes and bundles from the cars. I can't believe how much stuff we bring. When my family came in by float plane, we brought a minimum of supplies. Even after we could drive in, we were frugal, just the necessities. Now, we seem to bring everything that we can think of, on the off chance that it might be needed. The children race upstairs to choose their bunks. Each spring, they have raucous negotiations around the assignment of beds. The most prized are the furthest from the door, closest to the windows. Roger and Fin claim bunks even though they'll spend the summer months in the boathouse. They still like the warmth of the bunk room in the spring and fall. Bronwen gets there first and claims a prime top bunk with a reading light, then clatters back down the stairs in her bathing suit, hotly pursued by Roger.

"I'm going to be the first one in the lake this year," they shout in unison as they jostle each other through the front door.

"Not if I can help it," says Kevin, who strips down on the porch and runs after them.

The three of them race to the beach and jump off the dock into the water. They leap in and out like a fiddler's elbow, as Gran would say.

"Yikes! It's freezing!"

"It gets colder every year!"

I wrap them in towels as they come through the door, their skin icy cold. Hugging the towel close, they dash back upstairs to pull on the warmest clothes they can find.

I stand in the sudden peace and look out through the big living-room window at the view that stretches to the south before me. The water funnels through steep-sided mountains to a forested peak at the far end. The furthest peaks that I can see, just across the border, south of us, in the United States, are covered in glaciers. The late afternoon light washes the mountains with a blush of rose. Wisps of pink stroke the clouds above the lake, the sky a dusty blue fading to ash. The water is calm; occasional gusts of wind weave waffle patterns across the surface. The water mirrors the sky and the mountains, magnifying their beauty, framed in my window. I take a deep breath. I can't imagine my life without this.

Kevin comes up behind me and gives me a hug.

"Beautiful, eh Ma?" he says, his eyes following mine down the lake.

Although at 1.8 metres he now towers over me, he resembles me, with his fair complexion and affable nature. Michael, two years younger, is slightly taller than Kevin, a carbon copy of Patrick in the way he walks, his closely-cropped hair and gregariousness. The similarities continued into their chosen professions: Kevin followed me into law; Michael followed Patrick into executive recruiting.

I hear Michael behind me listing off the jobs that need doing and lining up the kids. The family is huddled around the island in the kitchen. We've done this opening-up so often that everyone goes to work on his or her allotted job. I direct where the provisions go that have been unloaded from the cars. Patrick asks the boys to start the fires that will drive off the winter's chill that lingers inside the house. Their coats still on, Fin and Roger begin with the wood stove in the kitchen, then the fireplace in the living room. Will, as always, wants to help.

"Okay, Will, bring the kindling," says Fin. "You can crinkle up the newspaper for us."

A smell of sulphur as the match is struck, the crackle of kindling and a waft of woodsmoke. The boys' fires create beacons of heat in the cold cabin, even though the logs won't absorb the warmth of the fires for several hours. Patrick supervises, piling larger logs into the fireplace, and using the bellows to draw the flames higher.

"Okay," he says. "The fire's really going now, boys. Let's get to work on the water."

Kevin and Michael gather the canvas bag holding wrenches and pipefittings and sling shovels across their shoulders; one of them carries the filter box that attaches to the water pipe. Patrick calls to Fin to bring the hipwaders. They head up the trail to Cupola Creek where we connect our water line. The rushing stream flows from the mountain ridge high above my cabin with froth and tumult, making a steep descent of 900 metres to Chilliwack Lake. Meltwater from snow that stays on the peaks nearly year-round. We call my cabin Creekside because it is the closest to Cupola Creek. We sleep to the sound of rushing water all year except in late August when the creek dwindles to a trickle.

About a third of the way up the mountainside above our cabin is a bald outcropping of rock my grandfather named the "Cupola" because it looks like the dome of a church. Poppy was so impressed by this rock hovering above his cabin that he named the cabin "Cupola Lodge," and the creek that feeds our waterline "Cupola Creek", and the company he formed to own the land "Cupola Estates."

The sounds of the men of the family—Michael wrestling with the boys, the dogs barking, Kevin calling to everyone to keep up—fade up the mountainside. Usually, I carry on with chores in the house, but this time I follow after them. I feel the pull of nostalgia, connecting the water, something I used to do with my grandfather and my dad. The trail winds among the trees, around a Douglas fir upended by a winter storm, its roots yanked out of the soil and reaching as high as the boys' heads, then through a grove of cedars rimmed by ferns.

When the creek appears, the trail follows its banks for another hundred metres to the holding tank. The bright green polyethylene fifteen-hundred-litre tank sits on a wood platform next to the creek. A length of black plastic pipe connected to the holding tank is looped and tied with rope in the tree beside the tank. Patrick unties the pipe while Kevin and Michael scramble up the rocks beside the creek ten metres beyond the tank. Here, we've dammed the creek to create a small pool. Every year it needs rebuilding before we connect the water system. Kevin takes off his favourite blue plaid lumberman's jacket with a hole in the back where a mouse chewed through it and puts on the hipwaders. He steps into the pool.

"Lucky for me you like cold water," says Michael, laughing at his brother, who grimaces as he wades into the icy stream.

They both work with shovels, digging the pool as deep as possible, their feet and hands growing numb in the cold water. The boys pass them rocks to raise the dam around the pool. Once the pool is deep enough, Michael places the pool end of the pipe into the filter. He puts the filter and pipe in the pool and secures it underwater with more rocks, then he signals to Patrick to open the tap at the water tank. They pause, listening for the sound of water flowing from the pool into the tank. Patrick gives a thumbs-down. "Must be an airlock," he calls.

Kevin and Michael clamber from the pool to the tank, shifting sections of the water pipe up and down until eventually the airlock is released and water starts to fill the tank. The men high-five all round.

"Okay, Ma, you can close the taps now at the house," Kevin says. I hurry back to the cabin to close the taps left open when we drained the system last fall.

The men gather up the tools and follow me back down trail. On the way down, they'll open the valves that allow water to flow to each of the three family cabins, Creekside, Cupola and Sandpiper. Back at our place, Kevin and Michael check the plumbing, hoping not to see the tell-tale spray of a leak.

"No leaks this year," says Michael.

"Yeah, remember last year, when the pipe under the sink sprayed all over the bathroom?" says Kevin.

"No fun mopping that up."

My daughters-in-law, Joanna and Molly, are busy cleaning the kitchen cupboards, whisking away all evidence of mice.

"Oh, oh," Molly says, "they got into the cookbook drawer. Look at the pancake recipe. It's all chewed up."

Sure enough, there's a nest of shredded paper in one corner and mouse droppings scattered everywhere. Molly takes everything out, turns the drawer upside down, disinfects it, then puts back the cookbooks.

"We need to set up our mouse catcher pail," says Patrick.

"Can I do it?" Will asks, and Patrick kneels beside his youngest grandson to show him how to punch a nail through both ends of an empty pop can. He threads a straightened coat hanger through each end. Together they fill a pail half-full of water and set the wire across the open mouth of the pail. The pop can twirls easily on the wire.

"Don't forget the peanut butter on the can!" says Will. Patrick smiles, enjoying Will's enthusiasm, and grabs the jar from the cupboard. Will slathers the can with peanut butter.

"Good man. You'll be a champion mouse-catcher yet."

"Where's the record of how many mice we caught last year?" asks Roger.

When he was five, Roger told his kindergarten class that his grandfather drowned mice and cooked them in the oven. His teacher called to ask Roger's mother if this "show and tell" was true.

"Well, sort of," Molly said. "His grandfather catches mice in a pail, then burns the dead mice in the wood stove."

"Let's hope we catch the mouse that chewed up the pancake recipe," Molly says now.

"You know the recipe by heart," Joanna says to me. "Can you rewrite it?"

I can. "Anna's Pancakes" is a recipe from my mother given to her by Anna, a mother's helper who came to Canada from Sweden in the 1950s. Anna eventually married and moved to Prince George, but her pancakes have lived on, prepared every Sunday for breakfast at the cabin. I made them for years, then Patrick did. Now Kevin has taken over the tradition.

Nora lines up the flashlights on the dining table and inserts the batteries she took out last fall. She flicks the flashlights on and off, checking to make sure each one works. Her intermittent flashes of light play a Morse code on the log walls. Bronwen and Fiona set out the candles: tealights for the votives on the table, tapers for the mantle over the fireplace. The three girls help me wipe the surfaces in the cabin with a damp cloth. The log walls have acquired a film of dust over the winter. Fiona gets out the speaker, connects her iPhone and Taylor Swift fills the room. The girls sing along, dusting in time to the music.

I take down the old black-and-white photos that hang on the log wall by the stairs to dust them off. The girls and I take a break from cleaning and cozy up on the couch in front of the fire to look at the pictures.

"Okay, ten minutes, then back to work," I say with mock sternness. These breaks with the grandkids are my favourite part of cabin-opening day.

The first photo is of my grandfather, Poppy, paddling with Gran in the old dugout canoe. Christopher and Frances Webb: the family's pioneers, the ones who first brought us here to Chilliwack Lake. The picture must have been taken in the late 1920s; on the shore behind them is the original one-room log cabin. We pull over a photo of my mother, Frances, when she was about twenty, standing on the pontoon of the float plane that flew her and Gran and Poppy in to the lake.

"Noni's beautiful," says Nora. As a toddler, Kevin named my mother. We were in the car and Mother was complaining that she had "no name" yet. The two-year old in the back seat started chiming "Noni, Noni."

Nora and I peer at the photo. Mother really is beautiful, a tall, willowy young woman with wavy chestnut hair, a Katharine Hepburn look-alike. The last picture is of me with my brother Chris. We're holding a stick between us with two rainbow trout hanging by their gills. Chris, with a mop of brown hair hanging in his eyes and a wisp of a smile; me standing awkwardly in cut-off pants, my dusty blonde hair cut bowl-shaped and ragged. Both of us are barefoot. The only family member not represented on this wall of pictures is Uncle Rowland, my mother's older brother. His place in the family is complicated. I don't mention his absence to the girls, but I notice it. I always do.

"Back to work, girls."

More cleaning, then we make up the beds, hang the towels, scrub the bathrooms. Roger and Fin fill the firewood and kindling boxes. Everyone helps empty the storage boxes. We've had break-ins in the past and now we box up and hide away our special treasures. Roger, who has an uncanny eye for detail, helps his grandfather place the treasures through the cabin. The collection of guest books, one for each year since we built Creekside in 1989, is arranged on the bookcase by the fireplace. The pair of carved wooden ducks, our first housewarming gift, is placed in the centre of the mantle. The yak bell from our trip to Nepal is hung by the fire. The tortoise shell from our teaching days in Zambia finds its place by the wood stove.

By dinnertime, most of the indoor jobs are done. We eat, then relax by the fire. We haven't yet got out the jigsaw puzzle and no one feels like playing cards, so we sit and chat. The light flickers across the children's faces.

The next morning, after a breakfast of Kevin's pancakes, we tackle the outside jobs. The beach is strewn with debris dragged in by winter storms. The kids haul the branches into a pile near the shore, ready for our annual inaugural bonfire. Fin and Roger retrieve pails from the shed that we fill with water to rinse sand from our feet and put them at the bottom of the stairs to the porch. The lake is choppy with the wind blowing onshore, whitecaps

The author's mother, age nineteen, wearing a sweater made by the Cowichan knitters of Vancouver Island. Her sweater has been handed down through three generations. Courtesy of Webb family.

visible halfway down its length. Waves slap against the dock, which, I note happily, has survived the winter and looks in good shape. I look out past the dock to the deadhead with boards nailed up one side as a ladder. In the summer, the kids climb this home-made ladder and do crazy jumps off the top. Repairs are needed: a couple of the slats slant sideways and one is missing altogether.

The deadhead is the top of a tree that fell and lodged in the bottom of the lake a short distance off the beach. The trunk rises about six feet from the water into the air, on a slight tilt. I don't know how the tree came to rest here but we have several such dead-heads at our end of the lake. I cannot remember a time without this dead-head. It shows up in every photo, in-cluding the grainy black-and-white photos from the 1920s and 1930s. It pulls my eye down the lake to the mountain peaks at the south end. Dark clouds cover the sky. The water reflects a silhouette shrouded in grey.

I shiver and pull my mother's Cowichan sweater more closely around me. It has a tree and deer mo-tif and bands of cream, grey-brown and dark brown, with a zipper down the front and a shawl collar. It is sur-prisingly water- and wind-proof. I like having Mother's and Gran's old Cowichan sweaters, a reminder of our early days here and a hint at In-digenous history in southern British Columbia. Nora is wearing Gran's.

Kevin and Michael are wearing their own Cowichan sweaters given to them by my Dad.

"Look, Rani. The wool on my sweater is unravelling. Here at the sleeve," says Nora.

"Yours has a hole too, Rani," says Bronwen.

"Who made these sweaters?" asks Fiona.

"The Cowichan people."

"From around here?" asks Bronwen.

"No, from Vancouver Island."

"How come we have them?"

"I don't know. I think Gran's is about eighty years old."

"I like wearing them," says Fiona. "They're so warm."

"Maybe I should try and get them fixed," I say, although I have no idea who would do such a thing, or how I would go about finding them. Maybe someone in Sardis will know.

One by one, the boats are brought out from the basement. We used to store the boats on sawhorses in the boathouse that sits at the edge of the beach in front of the dock. Each year, once the boats were taken out, beds would be set up converting the boathouse to a sleeping cabin.

The boathouse has a rustic charm. Hand-split cedar boards create rough walls and floors that mice and chipmunks can run through with abandon. Kevin and Michael unlock the wooden shutters that protect the windows during the winter and secure them against the walls. Fin and Roger have already staked out the boathouse as their summer quarters. The prospect of sharing the boathouse with mice, squirrels and bats doesn't bother them in the least. They like the independence. And they can leave the place as messy as they like. The girls are happy, too: more room for them in the bunk room. Only Will has his nose out of joint, still forced to bunk with the girls. I expect he'll be pushing his parents to let him sleep in the boathouse this summer, too.

We bring out the wooden rowboat first, its gunwales, inside ribs and seats a clear wood varnish, its outside painted white. Oars and oarlocks are found and placed in the boat. The shamrock green fiberglass canoe is next, its hull a cross-hatch of white scrapes, evidence of the rocky shores.

The dugout canoe comes last, a craft that has been with us since the late 1920s, so long that it seems like part of the family. My earliest memory is of paddling in the dugout canoe with Poppy. I have a recollection of Poppy telling me that "Indian Billy" carved this canoe and gave it to Poppy in payment for a debt. I wonder if that's true. And what was the debt?

Four of us are needed to carry the heavy dugout down to the beach. Seeing the canoe resting on the sand in front of our cabin completes the official opening, but I have one more private ritual.

The sky threatens and a stiff wind blows, but I step in, careful to keep my balance.

"Give me a push, will you, Michael?" And my son obliges.

My paddle cuts into the water. An occasional wave hits hard and sprays my face. This early in the spring, the water is so cold that I wouldn't last long if the canoe tipped.

I struggle to keep the canoe on a straight course. I can hear my Poppy's voice, "Always stay close to shore when the wind is up."

And I do. A hundred metres down the shore, I turn around and head back to the cabin. Only a short trip, but I will have many more solitary paddles this summer. This first one though, just me and the dugout, feels like coming home.

Warmth greets me as I open the kitchen door. I close it quickly, aware of the chill of the wind sweeping in behind me. Molly and Joanna are making grilled cheese sandwiches and tomato soup, perfect for a blustery day.

After lunch, I get out the summer jigsaw puzzle. We have shelves in the basement stacked with puzzles, one-hundred-piece puzzles for the children when they were little, a thousand pieces for us now. We do at least three each summer. Often, we have both a kids' puzzle and an adults' puzzle going on two different boards set up on tables by the front window with the view down the lake.

"Roger, come help me choose our puzzle," I say, and we head downstairs. Roger is the puzzle master, with Will following in his footsteps.

"Which one do we want to start with, the surfboards on the beach or the sailboats?"

"Let's do the surfboards," says Roger.

We bring the box upstairs. Roger sets the puzzle board on the table. We pull up two chairs, open the box and spill out the pieces. We have been doing puzzles at the cabin since Kevin and Michael were little. I buy some new ones each year, but we repeat the old ones too. Some of them we've done twenty times at least. They never get any easier.

"Okay," says Roger, "I'll pick out the edge pieces."

"I want to help," says Will.

"Me, too," Fin chimes in.

For a while, the four of us focus on putting together the outside edge.

"I see a couple of hummingbirds, Rani," says Fiona. "Let's put out the feeders."

I put on the kettle and Fiona retrieves the hummingbird feeders from the cupboard. The kettle sings out. Together, we mix the nectar, one part sugar to four parts water, Gran's formula passed on to a new generation. We fill the feeders, letting them cool before we hang them from the porch rafters. Fiona and I watch at the window. She leans into me, wrapping her arms around me, our heads nearly at the same level. I love these impromptu hugs. Within minutes, a hummingbird darts in and hovers at the feeder, our usual rufous, a glimpse of the red throat of the male flashing from under his brown head and side feathers. Soon others join him at the feeders, feisty and aggressive. They screech and dive, fighting to get the best perch on the feeders, a blur of wings and flashes of red. Their loud wing buzz and high-pitched staccato call pierce the air.

"Now can we mark our heights on the wall?" asks Nora. She never lets go of an idea. As the oldest, she remembers all the traditions and insists on following every one.

"Okay," I say, "go get the others."

Nora calls in the boys from outside and they come running to the bathroom. On the wall beside the sink is a board with the measurements of the children marked in pencil dating from 2003, when Nora was three and Bronwen and Fiona just two.

We start with the youngest. Will squirms as I take a ruler and mark his height with a pencil line. Each one takes their turn, commenting how much they have grown since last summer. Nora goes last, and, as I expected, she has grown the most, nearly four inches, followed closely by Fin. And yes, Nora has eclipsed my five feet four inches by nearly one whole inch.

"I knew it," Nora says, running to tell her parents and Patrick.

"How about some tea?" I say to the others.

"And cinnamon toast?" they say in unison.

This is part of the ritual, too. It's as if we're all singing a favourite song. No one misses a verse. "Sure," I say.

We sit around the coffee table in front of the fire, sipping our Earl Grey and munching squares of toast. Bronwen arches her eyebrow at me and points at her brother. Will's elbows are on the table.

"Hmm, Will, elbows on the table?" I say. Will quickly takes them off and gives me a sheepish smile.

"Oh, oh," says Nora, and the rest of the kids chime in, "Rani can never take you to have tea with the Queen!" This started when the kids were little, my nudge towards good manners, and the phrase has stuck.

We settle back on the couch, finishing off our tea. The cabin is officially opened, in all the same ways it has been opened every spring for nearly a century. There is nothing extraordinary in what we've done except that we keep doing the same thing, year after year, the ritual settling into our bones, another kind of genetic code. All twelve of us are sprawled on chairs and sofas, on cushions on the floor, quietly contemplative, staring at the fire's flames.

"I have some places I want to explore with you this summer," I say to the grandkids, laying out my desire to learn the history of the lake to preserve the story for them and future generations. "I've heard that there's a gravestone of an American soldier who drowned in the lake a long time ago."

"Where?" asks Roger.

"Down at the south end, near Depot Creek."

Fin perks up. "Can we go to search for that guy's grave?"

"Yeah, me too," says Will. "Don't go without me."

"Sure, we'll all go," I say. "At the beginning of the summer. When we're back here together."

"How will we find it?" asks Nora.

"Well, I'll go to the archives in Chilliwack and see what I can learn."

"Who is this soldier? Why was he here?" asks Bronwen.

"Will his bones still be there?" says Will.

"I don't know."

"Can Dad come with us?" asks Fiona.

"Oh yeah, Kev and I wouldn't miss this," says Michael.

"Can't be an expedition without all of us, right?" says Kevin.

"Okay, Rani," says Nora, "you have to figure this out so we can go find the grave."

"I'll go to the archives next week," I say. "I'll see what I can find out."

My mind is racing now. How will I find out where the grave is? The kids are so excited that I better get going on this research right away. I don't like not knowing the answers when the kids ask me questions. I go over to my desk and get out the notebook I bought to record my summer's quest. Its dark green leather cover reminds me of the forest's reflection on the water. I open to the first page and start my list of questions.

Who was this soldier?

Where is his grave?

GEOLOGY AND GEOGRAPHY

"How deep is the lake?" Adrian asks. Adrian and his fiancé, Becky, are visiting Chilliwack Lake for a few days in early May at the end of their trip across Canada, a reward for finishing their doctorates in geology at the University of Oxford. Patrick and I are here for the weekend hosting Adrian and Becky. I won't move out to the cabin for the summer for another month.

"No idea," Patrick says.

"We've always wondered," I say.

"We can figure this out, can't we, Adrian?" says Becky.

"Really? That'd be amazing," I say.

Adrian has been to the lake twice before, once as a boy of ten with his parents and again ten years later, when he spent the summer helping Michael stain our brand new log cabin.

I've known Adrian since he was born. Adrian and Kevin were born at the same missionary hospital, a year apart. Patrick and I met his mother and father in 1970 while we were teaching at Namwala Secondary School, a boarding school in southern Zambia. Joy and Gordon Foster were our mentors—they'd been teaching at the school for a year when we arrived—and we became close friends, keeping in touch after we returned to Canada and they went back to northern England.

I'm twenty-two when we meet. Joy is a few years older and my role model as I make my first foray into teaching. The school is coeducational but the students in the advanced classes of each grade are all male. The first week of school, I present myself to my English class, Form 3E, all boys, many older than I am. Here I am, a newlywed with long blonde hair and a miniskirt who looks like a teenager herself. The students ignore me. I ask for order. The noise continues. I don't know what to do. My voice quivering, I tell the class that I'm leaving and going to the staff room. When they want to listen to what I have to teach them, they can come and get me.

When I get to the staff room, Patrick takes one look at my face, turns me around and takes me for a walk, letting me talk my way through.

A half hour later, a student comes to fetch me. "Madam, we are asking you to come back and teach us," he says. Joy squeezes my arm and Patrick says under his breath, "You can do this."

Heart pounding, knees shaking, I walk back across the compound.

Voices hush as I enter. Then a scraping of chairs and all the students stand up.

"You are the lucky ones, you have a place in school. You are also lucky to have me. I can teach you what you need to do well in your exams." I pause. Complete silence. "I expect respect. I will give you respect in return."

I take a deep breath and start the lesson.

This class becomes the leaders of the school. They also become the students I most hate to leave.

It's at Namwala Secondary School that Patrick and I meet Simon and Lydia Maonde. Simon, a Zambian, is the headmaster. He rules the school with an iron fist. The students are scared to death of him. So am I. His wife, Lydia, is a teacher and administrator of one of the four girls' dormitories. She's also the mother of four children, the youngest of whom is two. She's daunting. She has to be: she was one of the first young Zambian women to graduate from secondary school and go on to university. The two of them make a searing impression on Patrick and me.

Forty-two years later, we received an appeal from the Maondes. They retired to a community that had no school and so they started a school in their home. They were teaching two hundred children in the bedrooms of their house in two shifts. Now in their eighties, they asked for help to build a school as their legacy to the community.

When this letter landed on my desk, I was winding down my law practice and looking for a challenge. I was casting my thoughts towards local charities. In an instant, I decided that I would do this instead. I flew to Zambia. I had to see whether Simon and Lydia were the visionaries that I remembered. Could I believe in them enough to put my heart and soul into this project? After two weeks with them, I was convinced. With other former teachers, I helped establish a Canadian charity, Friends for Zambia Society, and started raising money. Over the next six years, we funded the building of eleven classrooms, a water well, washrooms, administration offices and a library. Every year, I visit the Twitti School in Zambia, the thread of my life now woven into the fabric of the Maonde family. And I received an unexpected gift in Simon. The man who terrified me as a young volunteer became my beloved African father. We worked hand in hand. My greatest joy was to help him achieve his dream.

Having Adrian here brings Zambia alive for me again. Patrick and I reminisce about watching Adrian in our front yard digging in the dirt with his trucks.

"What do you need to get started?" I ask, pulling us back to the dilemma of the depth of the lake.

"Do you have any maps?" asks Adrian.

We rummage until we find some topographic maps. Adrian and Becky spread them on the dining room table.

"You can see that the water in this valley flows from south to north," says Adrian, studying the maps.

"Yeah, look how the Little Chilliwack River flows from the mountains south of the United States border, then widens to form the lake, then the Chilliwack River flows out at the north end," says Becky. "Somehow water was trapped here to form the lake."

"I've never noticed that before... fascinating."

The maps range in dates from 1975 to 1993. Adrian and Becky decide to use the most recent, labelled *Fraser Valley Scale 1:100,000*.

"That's from over twenty years ago. Not very up-to-date," Patrick says. "Is that okay?"

"Oh, sure. It's fine," says Becky. "You don't have any graph paper, do you?"

To our surprise, we find an old pad at the bottom of a desk drawer. Out comes a protractor and calculator from Becky's packsack. The map confirms that the surface of the lake sits at 600 metres above sea level. Adrian and Becky jot down the elevations of the surrounding mountains. On the west, from north to south, Mount Webb rises to 2,163 metres; Macdonald Peak, 2,194 metres; and Mount Lindeman, 2,309 metres. On the east, from south to north, sits Paleface Mountain at 1,770 metres, Mount Meroniuk at 1,789 metres and Mount Edgar at 1,987 metres. The mountain peaks are identified with dots on the map. Adrian and Becky measure the distance across the lake between the peaks.

Soon, the table is covered with graph paper scribbled with computations. Over lunch in the sun on the porch, they talk about the uncertainties they face in making their calculations.

"We don't know the gradient of the mountainsides," Becky says.

"Or whether it is consistent," Adrian adds. "So, we'll have to make some assumptions."

"We also need to see the river outlet," says Becky.

After lunch, they drive to the provincial park where they take a trail that leads to where the Chilliwack River begins in the northwest corner of the lake. This is the outlet. From there, they hike a kilometre or so along the bank of the river. They come to a bluff that falls steeply to the valley below. They reappear a couple of hours later back at the cabin.

"How'd it go? Did you find what you're looking for?" I ask.

"Yes. We came to a bluff where we could actually see out into the valley below," says Adrian. "The north end of the lake—this end—looks like it's been dammed."

"We still have more work to do," says Becky, "but we're getting close."

We leave them to their mathematical calculations, full of questions but not wanting to interrupt.

At about five in the afternoon, they emerge onto the porch where we are washing off the outdoor furniture.

"We figured it out," says Adrian. "The lake is one thousand feet deep."

"Wow. Amazing!" Patrick says. "How'd you figure that out?"

They produce two cross-sections on graph paper, one across the southern end and one across our end of the lake.

"Okay, we know the surface of the lake sits at 2,000 feet," explains Adrian. "It's roughly seven miles long and one mile wide. The mountains surrounding the lake range from nearly 6,000 feet to 7,500 feet and fall steeply into the water. We made the assumption that the mountainsides have a constant gradient."

Becky interrupts, "This end of the lake is unique. Originally, there was no lake, just a continuous valley covered by a glacier. The receding glacier deposited sediment that created a dam about a half mile thick at this end."

"That's what we call the 'benchland,'" I say.

"Right. The benchland is the dam that's holding back the water and creating the lake. Your cabins, the MacLeods' and the provincial park are on the benchland."

"You guys are incredible!" I say. "We've lived here almost a century and never knew any of this. Can you put your notes in the guest book? Then we'll always know who figured this out."

On one side of the page, they glue their drawings of the cross-sections of the lake and, on the facing page, write and sign their conclusion:

> The two sections show a depth to bedrock at more or less 1,200 ft below the lake. However, lake bottom filled with sediments deposited by glaciers. Probably 200–300 ft of sediment, therefore lake depth is more or less 1,000 ft.

> Dr. Adrian Foster BA, MA, PhD FGS
> Dr. Rebecca Buckley BA, MA, PhD FGS

Patrick and I started keeping a yearly guest book in 1985 when we moved back to Vancouver after living away for nearly twenty years. We wanted to record not only the names of our guests but also highlights of the summer, noteworthy events like opening the cabin, putting in the water, the arrival of the first hummingbird, the first person to swim in the lake each year. We took Polaroids and pasted them into the book, too. Guests added poems and songs. "Ode to Patrick," penned by one of the boys' friends, bemoans his hard work on the woodpile from dawn to dusk with barely a moment to water ski!

We have Gran and Poppy's cracked, green leather guest book from the 1940s, the pages discoloured on the edges. Inside are columns marked Date, Name and Comment. The comments are sparse and brief: "Frances's apple pie was delicious" or "Thank you for your hospitality." For some years, no record exists.

Adrian goes to the bookcase at the back of the living room and picks out the guest book for 1990. Becky laughs at the photo of him, mop of brown hair in his eyes, glasses perched on his nose. "Look how skinny you are," she says.

> After a brief reign as the 'Cappucino King' and a longer one as Mike's work-mate, it's time to go on with the holiday. Despite cramps from painting and vertigo-inducing ladder positions, the work was really quite fun. The skiing was fantastic—dock start next time!

"It doesn't seem that long ago," says Adrian. "Can we go for a boat ride tomorrow before we have to leave? I'd like to show Becky the south end of the lake."

"Sure. Michael brought the motorboat in last weekend, so we're ready to go. Let's go in the morning when it's calm."

The next morning, although the sun is out, the early May breeze off the water is cool. The snow line is still halfway down the mountains. I canoe Patrick out to the boat tethered to its orange buoy in the bay in front of the cabin. He takes off the cover, folds and stores it away. Adrian and Becky bring jackets in case the wind comes up and carry the cooler with lemonade to the dock. Patrick stows the gear, checks to make sure the life jackets are in the compartment in the bow. The motorboat is a recent addition, a Campion bowrider, with seating in the bow as well as the stern of the boat.

"Nice boat," says Adrian to Patrick. "New?"

"Yeah, large enough for the whole family, although it's pretty crowded with the twelve of us. Why don't you and Becky sit up front?"

We step on board and get settled in our seats, Patrick in the captain's chair, me beside him and Adrian and Becky up front. It's sunny, with just a slight breeze. The sun reflects pinpricks of light off the ruffled water.

"What a day for a boat ride," says Becky, her dark curly hair unruly in the gust of wind as the boat starts up.

Our plan is to take our usual tour around the lake. For reasons unknown to any of us, we always motor counter-clockwise, across the north end, down the west side towards the mountains in the south, then home along the eastern shore to our cabin at the north end. Patrick steers the boat out fifteen metres from the shoreline, where the water darkens from pale emerald to black.

"Becky," Adrian says, "look over the edge on this side of the boat. You can see nearly every grain of sand."

"Now, look on the other side."

"Wow. It's nothing but black. I guess the sharp drop is the edge of the benchland."

I'm grateful we can't see to the bottom. For years, my grandparents, my parents, and I too dumped our sinkable scrap in the lake. That is how we got rid of anything that we couldn't burn or compost. Tin cans, metal bedsteads, an old cook stove all went into the water. I wouldn't want to face all that garbage, but even so, what remnants of history must be sitting down there!

No one considered the environment when I was a girl. Now we do. We sort metal, plastic and glass to take to a recycling depot in town. Will and I often spend our morning paddle picking up litter along the shores of the lake. We wouldn't dream of dumping a bottle cap, let alone a wood stove.

Heading west in the boat, we pass Gran and Poppy's Cupola Lodge, built in 1924, and Mother and Dad's Sandpiper Pan-Abode, built in 1963, each set in its own clearing. (Both of these cabins are now used by my brother Chris and his family.) Another kilometre of beach, interspersed with large rocks and cottonwood and alder trees on the shore's edge, and the MacLeod place comes into view. In 1939, the MacLeods bought a parcel of our land from Poppy and, the following year, built their log cabin. They have been our only permanent summer neighbours ever since.

"Wow. It's warm. I don't need this jacket," says Becky as she strips off her windbreaker.

Patrick is cruising along slowly and the sun beats down on the four of us. I take out my sunglasses and put on my cap. We chat about Adrian and Becky's next destination: Banff and the Rocky Mountains. It's easy to talk above the low rumble of the motor.

Beside the MacLeods' cabin is the boat ramp and swimming beach for the provincial park. The park was created in 1971, encompassing a four-hectare parcel on top of the bluff at the edge of the lake, with a handful of rough backcountry camping spots. Patrick and I returned from Zambia a year later to find a public campground practically on our doorstep. The public beach felt far too close even though we couldn't see it unless we went to the end of our dock. I walked down the beach steeling myself to what I would see. Campers launched boats, picnickers laid out provisions on blankets, boom boxes blasted out music. My heart sank. I hated it.

It still rankles every time I go past. So much so that, whenever I go for a paddle, I always head east from our beach, the opposite direction.

After the boat launch, the shoreline is rocky with cottonwood, birch and alders leaning out over the water, and Douglas fir and western redcedar towering behind. We cruise up to the outlet of the Chilliwack River at the northwest corner. This is where the river begins, flowing north out of the lake in a smooth sweep, about fifteen metres across. The pull of the river is gentle here but it quickly narrows into rapids and travels a tumultuous forty kilometres to join the Fraser River on its way to the Pacific.

Crossing the outlet, Patrick concentrates on avoiding submerged logs, some sitting just below the surface, others rising a few feet above the surface. I catch Patrick's eye.

"Remember when the boys were small? You'd bring them here," I say.

"Fighting the Raspberry Tribe," Patrick laughs, turning to Adrian and Becky. "We used to paddle here in the canoe each morning. I started the game of watching out for the Raspberry Tribe. Kevin and Mike would attack the Raspberry Tribe from the logs at the outlet. I have no idea where the name came from."

"They'd come back to the cabin with tales of their adventures." Telling this story, I'm vaguely uncomfortable. The name has vestiges of the games of Cowboys and Indians I played as a girl. We never thought about such things thirty years ago, but now it seems disrespectful to the Indigenous people who occupied this land long before we did.

Patrick steers the boat south along the western shore. Dozens of creeks cascade down the mountainside, pushing water out over the rocky shore. The morning sun sets each tree branch in sharp relief against the forest behind. The Douglas firs are at least thirty metres high, with even taller western redcedars interspersed among them. I shiver as a thin bank of clouds draws a sudden curtain across the sun. We pull on our jackets, zipping up against the chill.

I search the tops of the cedar snags. I catch sight of a complex interweave of twigs and branches on a prominent snag, at least one and a half metres across and a metre high.

"An eagle's nest," I say pointing. The others turn to look. "Up the slope at about three o'clock. On top of that snag."

"I see it. Can we get closer?" asks Becky.

Patrick slows the motor. I reach into the glove compartment for the monocular, a single magnifying lens used by Patrick's dad during the Second World War. I pass it to Becky. A white head appears, then a couple of smaller tawny heads.

"I see some young ones," she says, passing the monocular to Adrian.

An adult eagle hops up to the edge of the nest, pauses, head swivelling, then lifts its wings and spirals up in slow circles over the water. The morning light catches its snowy head and tail feathers. It glides with outstretched feathered wings, then loops back to the nest.

We haven't always had eagles on the lake. Or maybe I didn't notice them before. Now, we see the eagles all summer long, circling and climbing high on the thermals rising off the water. We watch the offspring, mottled brown for a couple of years, until they too acquire the distinctive pearly head and tail. I am heartened to have them as neighbours. Every year, I look for the two nests on the western shore. I worry that the pressure of people, with their noise and pollution, will spoil the ecological balance. As I grow older, this concern has intensified. I want to protect this wilderness of ours, hold it in trust for future generations.

When I started practising law in 1987, the nascent field of environmental law caught my interest and I started up an environmental law group at my firm. Most of my clients were at first resistant to making environmental protection a priority. Logging road blockades in Clayoquot Sound and Meares Island had created a them-versus-us atmosphere. British Columbia became renowned for its environmental activists: Paul Watson of Greenpeace; Paul George of the Western Canada Wilderness Committee; and David Suzuki. We owe them a debt of gratitude. Without those environmental pioneers pushing the envelope, businessmen might never have opened their eyes. The value of good environmental practices became evident; industry leaders became converts. My own view was always clear. Environmental protection is not just a platitude; it is essential.

The return of the eagles each year strikes me as a good omen.

Patrick pushes the throttle forward. Wind blows our hair about in swirls. A third of the way down the western shore, he slows the boat again off a rock ledge that juts out over the water, creating a broad cliff face. The Jumping Rock.

"Mike and I used to come here after working all day on the cabin," Adrian tells Becky. "I'll never forget the shock of the cold water."

"We still come here in the summer with the grandkids," Patrick says.

A rock face on the western side of Chilliwack Lake has become a family favourite for cliff jumping. Courtesy of O'Callaghan family.

"The youngest can jump from the low end and the oldest at the high end. In August, when the water is low, the high end can be five metres above the water. Hey, Adrian, want to jump off for old time's sake?"

"Are you kidding?" says Adrian. We all laugh.

"Hold on to your caps," says Patrick. "We'll gun it to the south end of the lake." He presses the throttle and the boat leaps forward down the middle of the lake. We sit back in silence, the motor's engine drowning out any attempt at conversation. A multitude of creeks cascade down the mountains on both sides of us. When we reach the wide beach that marks the southern shore, he slows the boat to a crawl.

"Anyone thirsty? How about some lemonade?"

I bring out the plastic glasses and thermos and pour each of us a glass of lemonade. I pass around some cookies, oatmeal chocolate chip, made from my sister-in-law's recipe; she's the renowned cookie-maker in our family.

As we cruise along the shallows, we look down on schools of rainbow trout gathered at the mouth of the Little Chilliwack. Marshy, with meandering rivulets, the river flows into the lake from the United States through a long tapered valley. Only three metres wide at its mouth, the Little Chilliwack is little more than a stream compared to the wide rushing river that leaves at the north end of the lake.

"We come here for picnics in the summer," Patrick says, "but the mosquitoes drive us out as soon as the sun sets."

"I've heard a trail goes through an ecological reserve from here to the United States border," I say. "Cedars nine metres in circumference. I'd like to trek it some time this summer."

Geography has conspired to keep this end of the lake pristine. With no road access, little has disturbed it over the years.

Our mid-morning snack over, we pack up the cooler and set off again following the eastern shore. The wind picks up, bouncing the boat against the breakers. A creek spreads out through a promontory of sand. Depot Creek. We round the corner into a protected bay. A dock and some wooden buildings cluster in a clearing: a main lodge, several outbuildings, an A-frame that was once a chapel. Patrick slows the boat and lingers in the bay.

"This is the Church Camp," I say. "It's been here since 1946, two years before I was born. It was originally a camp for boys who'd lost their fathers in the war."

"What's it used for now?" asks Adrian.

"Local church groups from the valley. We don't see them much."

I remember, as a girl, watching the boys make their way on a flotilla of boats from our end of the lake to their camp on the eastern shore. I never got to know any of them, never exchanged a word. I eyed them surreptitiously. Although we were on the same pool of water, our paths rarely crossed.

Just past the Church Camp, Kokanee Creek tumbles into a sand estuary. I have to keep reminding myself that it's now listed as Paleface Creek on the topographic maps. I'll have to find out why the name changed. Another question to add to my notebook.

"The wind is really up now," says Patrick. "I think we'd better head straight for home." Adrian and Becky move to the back of the boat beside me and Patrick opens the throttle. The boat thumps down hard against each wave. We hunch our shoulders into the wind riding the bucking boat. With relief, we climb out at our dock.

"Thanks for the boat ride," Becky says warmly, "I've never seen anything like this."

"It's like the lake is frozen in time," says Adrian. "Pristine."

That's part of the pull of the lake for me: it doesn't change. But I have to admit that my life here *has* changed, from a brief summer sojourn, cut off from the rest of the world, to a three-season retreat with telephone and internet connection. From a one-room log cabin with an outhouse to a two-storey house with four bedrooms and hot showers.

The next morning, Patrick and I sit on the porch steps, bundled up in our fleeces. Adrian and Becky left yesterday after lunch: we are alone again.

"Wasn't that great to see Adrian and Becky?" I say, warming my hands around my mug of coffee.

"Yeah, and now we know how deep the lake is," says Patrick.

The heavy dew on the grass around the cabin and the chill off the snow-covered mountains reminds us that winter has not yet made its complete exit. We gaze out on the choppy water and chat about Adrian and Becky's findings. It is interesting to see our oasis from their fresh perspective. My own perspective seems to be changing daily. Our property is a prism through which I am looking at all of my actions. My retirement has provided me with the time and desire to dig more deeply into my relationship to this special place. I find I am asking questions. What is our responsibility to the environment? Should we be doing things differently? Are we doing anything that is harming the environment? I have read somewhere that even biodegradable soaps can contaminate lakes. I need to do more research on this. The lake is not just ours. What about the fish, the ducks, the otters?

Patrick points to something in the water. "I think the otter's back. Look out past the deadhead. See the ripple?"

I hurry inside to get our old Baker Standard binoculars.

"There, just to the right of the deadhead," he says.

"I see it. Definitely the otter. It's coming towards the dock."

We've had a pair of river otters here for as long as I can remember. They generally set up their home in the rocks just around the point in a bay we call Otter Cove. At this time of year, the male or female is out fishing alone, the mate staying with the kits back in the den. Their den is on land and has both an above-water approach and an underwater entry.

The otter slips up onto the end of the dock in a single fluid motion, its chocolate fur glistening in the early morning light. Through the binoculars, I can see its flat head and broad muzzle spiked with long silvery whiskers, its muscular neck and tapered tail. It stretches, rubs its belly, then its back across the planks of the dock. It pays no attention to the humans sitting several metres away on the porch. Silently, it slides back into the water and heads for Otter Cove, the essence of fluidity. First its head appears, then its sinewy back and tail curve wet and sleek out of the water and finally, with barely a ripple, it's underwater. Sometimes the otter will surface within nine metres; at other times, it'll stay under for thirty metres or more until we're sure it has disappeared entirely.

Seeing the otter for the first time this year is a landmark, an omen, like the eagles. It makes me believe that all is well. Am I deluding myself? I tend to ignore the things that I don't want to see. Maybe I am like an ostrich with its head in the sand?

Patrick and I linger on the porch. It is still too early to plant the flowers in the half-barrels that line the porch. We could tackle the myriad chores that await us every spring: setting up the tent; refinishing the deck chairs; clearing leaves from the eavestroughs. But instead we take in the stillness. Breathe in, breathe out. The quiet broken only by the whirring of humming-bird wings, the occasional caw of a raven. This is what the wilderness does for me. It slows me down and allows me to be still. It opens my senses. I feel the brush of the breeze on the hairs of my arm. I hear the rustle of dead leaves scattering across the beach. I breathe the fir-scented air. I feel alive in this moment.

I look up towards Mount Webb, its steep broad rock crown suspended over a glacier in an alpine amphitheatre. The glacier stretches across the scree above the tree line. Below it, a heavily forested slope descends precipitously to the western shore of the lake. The mountain was named for my grandfather, Christopher Everest Webb. The naming application was made by Uncle Rowly, my mother's brother, who documented the historical rationale:

> Mr. Webb first went to Chilliwack Lake in 1916 as an engineer with the Federal Water Resources. The original log lodge on the property was built shortly thereafter with the assistance of Charlie Lindeman, after whom Mt. Lindeman is named. For forty-odd years Mr. Webb's family have enjoyed and developed their property, Cupola Estates.

My grandmother saved Poppy's letters and gave them to me some time ago. I have brought this file to the cabin to start my research. I read Rowly's letter last night. I wonder now about the accuracy of the date. I've come across several letters that refer to Poppy's tenure at the lake. The date of first arrival varies, ranging from 1916 to 1925. The hundredth anniversary of our family's time at the lake is coming up soon. If we are going to celebrate, we need the correct date. Poppy was so meticulous, I'm surprised that his first visit isn't carved in stone. I go into the house and bring out my notebook. I add this to the list of questions.

What is the exact date of our arrival here? I go back in my mind to a photo of Gran and Poppy in the dugout in their early days at the lake.

Rowly's letter reveals something else. Despite a difficult father-son relationship, Rowly clearly loved his father. He wanted to give Poppy a gift that

The author's grandparents in a dugout canoe on Chilliwack Lake with their one-room log cabin in the background, circa 1926. Courtesy of Webb family.

he would cherish. Mount Webb was named before Poppy died. I remember how proud Poppy was of this achievement. And Rowly made it happen.

Roger's middle name is Webb. He is only eleven but already he feels a special connection to the Christopher Webb he never knew and to "his" mountain. He never tires of hearing about the time Kevin and Michael—his uncle and father—climbed Mount Webb when they were even younger than he is now.

Two blond heads are bent over the packs laid out on the porch. Kevin and Michael are rolling their extra socks, shirt and pants into the bottom of their packs to provide a cushion for the hard-edged flashlight. Sleeping bags are stuffed in, with rain jackets on top. Their water bottles are attached with carabiners to the side of the pack. Thick socks, hiking boots, long pants, long-sleeved shirt over a T-shirt, sweater, brimmed hat, packs on their backs, walking stick in hand: they are ready. The boys, eight and ten, spent most of the day before searching for the perfect walking sticks, whittling them to the right length and sanding them smooth. Patrick and I have the food, Coleman stove, tin plates and cups, aluminum pot and pan divided between us. We are lightly packed, foregoing a tent, planning to stay the night in the forest service cabin at Radium Lake at about the seven-kilometre mark. The essentials—map, compass, matches in waterproof container, Swiss Army knife and first aid kit—are stowed in an outside pouch. Also under the category "essential" are several Ziploc bags of Gorp, "Good Old Raisins & Peanuts," with the equally essential addition of chocolate chips.

The trail starts at the Chilliwack River, a kilometre in from its outlet, and follows Radium Creek up the valley just west of Mount Webb, a steep steady grade with relentless switchbacks, that rises 1,615 metres to reach the summit of Mount Webb. The boys are troopers. Buoyed by handfuls of Gorp and hiking songs, "She'll be Coming Round the Mountain" and "Jamaica Farewell," we keep a steady pace. As the day warms, we strip down to T-shirts. We have lunch on a large rock beside the creek, legs hanging over the edge, moss slipping off the rock and tumbling away with the current. Our standard hiking lunch, ham sandwiches, apples and Oreo cookies. We refill our water bottles with clear, cold water from the creek. After lunch, our pace slows until the sugar kicks in and we get our rhythm back.

We arrive at the forestry cabin by three in the afternoon. The tumble-down shack has a window with cracked panes of glass laced with cobwebs, a door that is crooked on its hinges, and roof through which thin slices of daylight can be seen. Inside the stove is missing its stovepipe and two sets of wooden bunks have broken slats.

"It stinks in here," says Michael.

The musty smell of rodents. We cut some cedar branches to sweep out the cabin and improve the smell of the place. Once we have a semblance of order, we go out exploring. The shelter is in an alpine meadow perched at the end of Radium Lake, more a pond than a lake, the ground marshy underfoot. I'll bet mosquitoes will be buzzing us later. Patrick and the boys head across the meadow down an incline and are lost from view. I follow the voices.

"Ma," Kevin calls out to me. "Come see this pool."

"Let's go for a swim," says Patrick.

Everyone strips down to underwear.

"It's freezing," shouts Michael, as he scrambles up the rocks to jump in again.

After a few dips in the frigid water, we lie on a boulder in the waning sun. As the shade creeps across our bodies, we start to shiver. We put our clothes back on and head to the cabin for dinner. Patrick gets a fire going in the fire pit outside while I add water to the freeze-dried "hearty chili" and heat it over the fire. After dried apricots and a couple of squares of Hershey's choc-olate, we sit by the fire, content to watch the flames and see who can spot the first star emerging out of the growing darkness. We unroll our sleeping bags on the dilapidated bunks, a single candle on a bench flickering shadows across the log walls. Here we are, in the wilderness, not in the safe and comfortable confines of our cabin. How do I feel, really? A tingle of exhilaration, this is exciting, we'll have stories to tell. A flash of fear, what dangers lurk in the darkness? A high-pitched buzz signals the arrival of the first mosquitoes. I bury my head in the sleeping bag.

"'Night, boys," I say, wondering if I'll hear the rustling of mice over the whine of mosquitoes.

A different sound sets my skin tingling. A crash. I sit upright.

"What was that?"

Another thump and clatter. Patrick gets out of bed and opens the door a crack.

"A bear's at the fire pit," he whispers.

"Will the bear come in?" Michael asks in a hushed voice.

We hold our breath, listening hard. After a few minutes of quiet, Patrick peeks through the door.

"He's gone," Patrick says. "Go back to sleep, boys."

After a fitful sleep, we fortify ourselves with hot oatmeal and tea laced with spoonfuls of sugar and set off to climb to the top of Mount Webb. The boys chatter about the bear, wondering where he went. Eventually, we fall silent as we clamber on hands and knees up the rocky slope.

"The view will be worth it," Patrick promises, egging the boys on.

And it is. We crest the ridge and suddenly I am like the eagle flying over-head, looking down on tiny, aquamarine lakes in the bowl below us. And further down the mountain is our lake, forest green and shaped like a tube sock. The white of our beach stands out against the dark water. The boys spot our cabin, a dot at the far end of the beach. I can see the wind playing across the water even from this height. The colour of the trees seeps into the water, shifting shades of green.

I turn around and look at Patrick sitting on the porch beside me. "I was just thinking about our hiking trip up Mount Webb with the boys back in 1982," I say to Patrick. "Did you know that Michael's thinking of taking Roger this summer?"

He gives me a smile. "I guess that won't be for a while, still lots of snow up there."

"Even Adrian and Becky said they'd like to do it next time they come."

"They're coming back?"

"I don't know. But everyone seems to come back eventually."

Friends who visit the lake always want to return. "This place is magic!" they write in the guest book. I know what they mean, but where does the magic come from? Is it the isolation? The peace and quiet? The pristine beauty? The fact that it never seems to change? What, exactly, is the magnetic pull of this wilderness lake?

I add that question to my list, too.

First Peoples

The days are stretching out as May turns to June, with warmth in the sun at midday. Snow is still massed on the far mountains across the border in the United States, but only patches remain on the peaks surrounding the lake. I am here on my own; Patrick works in the city through the week and comes on the weekend. The grandchildren are still in school, so the families visit sporadically on weekends, as their schedules allow.

My retirement provides me with the opportunity this year to spend three months at the lake from early June to mid-September. I have never spent more than a few weeks here and always with Patrick, or the boys when they were little. I have chosen to be at the lake so that I can do my research in the archives in Sardis, a village of twenty thousand, and in Chilliwack, a city of one hundred thousand. But even without the pressure of research, I would spend this time here. I miss my family during the week, but I savour the solitude. It gives me the chance to think about the influence of the lake on my life. At my age, I ponder the meaning of my life, what I want to leave as my legacy. I go for walks, exploring the far reaches of our property, wanting to discover every nook and cranny.

I walk along the beach, enjoying the burst of green on the branches of the cottonwoods and alders. Light dances on the water as the wind shifts and scatters. A shadow ahead snags my attention. Is that Eva Sepass? No, of course not. Those days are long gone.

I was just a girl when I'd see Eva walking our beach.

She comes towards me, a long stick in one hand that she sweeps across the sand. Occasionally she stops and peers down, cocking her head to one side. As she comes closer, I recognize her.

"Hello, Eva," I call out.

Eva Sepass of Chief Sepass's clan, that's what Poppy calls her.

She smiles and waves but continues her search, focused on the sand in front of her. She is scratching the surface. Every so often, she stoops, inspects the object before her and picks it up. She caresses it in her hands. When she reaches me, she passes an object, now warm from her touch, into my hands. It is small, no more than five centimetres in length, dark grey, rough-hewn and sharply triangular, with two notches at its base. An arrowhead, the once-sharp edges smoothed by time and the action of water and sand. I look up at her with a question in my eyes.

"Our people used these to hunt birds," Eva says to me.

"What kind of birds?"

"Ducks, geese, many other birds."

"Why?" I asked.

"To eat, of course," Eva laughs.

"Gran," I call out finally, because I must, "Eva's here."

Gran, white hair haloing her head, rimless glasses perched on her nose, comes out of the cabin shedding her apron on the porch and joins Eva on the bench in the shade of the birch tree. Gran and Eva share a pot of tea and gossip about the valley. Gran listens to Eva's stories, a warm smile on her face, as if there is nothing she'd rather do. I sit at their feet dreaming of tents in a clearing and canoes on the lake, the pungent smell of fish being smoked over the fire.

As a girl, I formed an idealized picture of life at the lake for Eva's people. I knew nothing of reality. When Dad, walking the beach near the campground, found arrowheads, he was intrigued, but that was all. The triangles of stone were curiosities, souvenirs he displayed on the mantle. How odd, I think now, that my father, scholarly and thoughtful, didn't ponder the connection of the first people to this land and to us.

Our cabins sit on land originally occupied by the Indigenous peoples. There were no treaties. The land was simply taken by the British Crown. We bought our land from the government of Canada in the 1920s and 1930s. At the time, no thought was given to the claims of Indigenous people. All that has changed. Now, government cannot sell land claimed by Indigenous people without consultation and consideration of their rights. Today, we wouldn't be able to buy this land. As I reflect now on my relationship with Indigenous people, I feel conflicted. This land originally was held by others. Possibly they lived here and would have loved it as much as I do. But this land is now ours. I would not want to give it up. I am ashamed to say that I speak from a position of ignorance. I know almost nothing about the people whose traditional territory includes the lake. I am deeply uncomfortable that neither I nor anyone in my family have taken the time to learn about their history, their connection to this place.

My journey of discovery, I realize, does not begin with my family. It starts with the people who were here first, long before we arrived.

Coqualeetza residential school, circa 1900. The school was built in 1894 by Methodist missionaries on land owned by the Skowkale First Nation. The property is now owned by the Stó:lō Nation. An adjacent building on the same property houses the Stó:lō Research and Resource Management Centre. Courtesy of the Chilliwack Museum and Archives, 2014.021.001.

Who were the first people to live at the lake?
What was their life like?
Why did they leave?
I come up from the beach and retrieve my notebook. I add these questions to the list. Tomorrow, I will drive to the Coqualeetza Cultural Education Centre in Sardis, forty-five kilometres down the Chilliwack River valley from Chilliwack Lake. The Coqualeetza Centre houses extensive archives in its Stó:lō Research and Resource Management Centre. This will be my starting point. The Coqualeetza Centre itself has a controversial history. In the 1880s, Methodist missionaries converted their home in Sardis to a boarding school for Indigenous children, one of the earliest residential schools in British Columbia. In 1891, the school was moved to an imposing three-storey brick structure with a gabled roofline located on Skowkale First Nation land.

Later, the building was converted into a tuberculosis sanitorium operated by the Department of Indian Affairs. The sanitorium was closed in 1969 and a controversy erupted as to the rightful ownership of the land. The Skowkale First Nation occupied the building and negotiated, successfully, with the government of Canada for the property to be given reserve status. In 1973, the Coqualeetza Cultural Education Centre was established in the former school/sanitorium, dedicated to preserving Stó:lō culture. Over time, historical documents were accumulated and, in 1996, the Stó:lō Archives were established. A new building, the Stó:lō Research and Resource Management Centre, was completed in 2010 to house the archives. The Co-qualeetza Centre continues to be housed in the original building.

I arrive at the Stó:lō Research Centre at eight in the morning, having arranged my visit with Tia Halstad, librarian of the Stó:lō Archives. Two four-and-a-half-metre figures carved from cedar—a man and a woman—stand as sentinels at the entry of the archives building. Traditional Coast Salish figures of greeting, they symbolize the welcome given to visitors, although the man keeps one arm down by his side, ready to protect his people and territory. In the archives, I introduce myself to Tia Halstad and Michelle Tang, her assistant.

"I am researching the early history of Chilliwack Lake," I say. "In particular, I want to know about the Indigenous people who lived in the area. Can you give me some guidance?"

"Oh, that would be the Ts'elxwéyeqw," says Tia. "Let me bring you the *Stó:lō Coast Salish Historical Atlas*. It's a good place to start."

"What about the life of Chief William Sepass?" I ask.

"I'll look into what we have," says Michelle. "Have you seen his book of poems, *Sepass Tales?*"

"Not yet. I've heard about it though. I'd like to buy my own copy."

"You could try The Book Man in Chilliwack. Or Longhouse Publishing in Mission. We also have some maps of early trails and trade routes through the Chilliwack Valley if that's of interest," says Tia.

"I want to see it all!"

They set me up at a small, round worktable and bustle about retrieving books and files. Pale light filters through the windows. I tug my sweater a little closer.

Tia lowers a stack of books on the table in front of me. The *Stó:lō Coast Salish Historical Atlas*, *The Chilliwack Story*, edited by Ron Denman of the Chilliwack Museum, and *Lower Fraser Indian Folktales*. A treasure-trove of information on the early history of the Stó:lō.

Before I start reading, I pause. This is the first time that I have handled archived material. I think about the people that wrote these diaries, drew these maps so many years ago. I am exhilarated and, at the same time, reverential. I am touching the past.

I am immediately captured by the story that unfolds before me. Ts'elx-wéyeqw is pronounced *Ch-ihl-kway-uhk*, anglicized as "Chilliwack." The

Ts'elxwéyeqw people, who lived on the shores of Chilliwack Lake up to the nineteenth century, traced their heritage from four legendary ancestors— Wileliq, Siyemches, Thelachiyatel, and Yexweylem—black bear brothers with white spots on their chests. Legend has it that the bears were transformed into men who became the leaders of the Ts'elxwéyeqw.

In the 1700s, the Ts'elxwéyeqw lived at the south end of Chilliwack Lake. They were connected with the Nooksak that lived along the Nooksak River to the southeast of the lake, in what is now the state of Washington in the United States. In those days, of course, there were no borders, no countries called Canada and the United States. No Indigenous maps have survived, either. Just endless, stateless, wilderness. At that time, the Ts'elxwéyeqw spoke a dialect of the Nooksak language. Their trading trails went from Chilliwack Lake south and east into Nooksak territory. The Ts'elxwéyeqw summered on the lake and travelled south to winter at lower elevations in the Nooksak watershed.

In the early 1800s, Chief Wileliq and his three brothers moved their main village to the north end of Chilliwack Lake near the river's outlet, just one-and-a-half kilometres down the beach from our cabins. In the *Stó:lō Coast Salish Historical Atlas*, I find a map showing the location of pit houses at the north end of the lake. I also find a drawing of what a pit house would have looked like. An oval hole was dug in the ground, two metres deep and eight metres in diameter. Posts and beams, often carved or painted, supported a roof structure made of saplings covered with earth to create an egg-shaped dwelling. People entered the house through a central opening in the top, climbing down a notched log ladder positioned near the hearth. Pit houses were used for winter and alpine habitation.

I am confused by the relationship between the Ts'elxwéyeqw and the Stó:lō Nation. I review my notes. The Ts'elxwéyeqw lived at Chilliwack Lake for thousands of years, well before contact with white settlers. The Ts'elxwéyeqw moved from the lake and settled in the Chilliwack River valley in the mid-1800s. Members of the Ts'elxwéyeqw people separated into nine First Nations, one of which is the Skowkale First Nation. This group, along with other Indigenous peoples in the Chilliwack area, joined together to form the Stó:lō Nation.

As I dig deeper, I struggle with the terminology: the difference between a "tribe" and a "nation." I read that generally, a tribe is a group of families who share the same ancestors and culture. A nation is a larger group of people that join together. The term "First Nation" came into common usage in Canada in the 1980s to indicate sovereign political status. Now I learn that the respectful term to use is Indigenous peoples. This is what I will use.

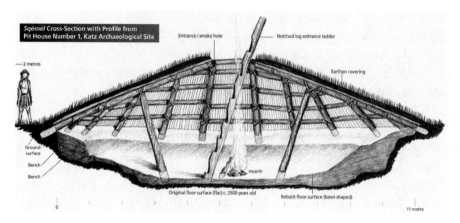

Sqémél Cross-Section with Profile from
Pit House Number 1, Katz Archaeological Site — Entrance / smoke hole — Notched log entrance ladder

— 2 metres
Earthen covering

Ground
surface
Bench
Bench
Hearth
Original floor surface (flat) c. 2500 years old
Rebuilt floor surface (bowl-shaped)
0
11 metre

A diagram from the *Stó:lō Coast Salish Historical Atlas*. Pit houses were built at the north end of the lake by the Ts'elxwéyeqw, the earliest recorded Indigenous people at Chilliwack Lake. Courtesy of Stó:lō Research and Resource Management Centre.

I type away on my laptop, taking notes, keeping track of sources and page numbers. I flag pages to copy so I can take critical information home with me. Unfortunately, the *Stó:lō Coast Salish Historical Atlas* is for reference only and cannot be checked out. The book is a jewellery box full of historical gems. I wonder whether I could find any remnants of pit houses at the lake. Not likely: such structures would have collapsed long ago.

This village with pit houses at the outlet was one of nine Ts'elxwéyeqw settlements along the Chilliwack River. An 1808 census shows the village at Chilliwack Lake with approximately four hundred and fifty people and the eight hamlets along the Chilliwack River with one hundred and eighty people each. The *Stó:lō Coast Salish Historical Atlas* presents an interpretation of historical Stó:lō demographics, acknowledging the uncertainty in establishing pre-contact population figures. Assuming that Indigenous populations experienced a 90 to 95 per cent decline over the first century of European contact, the pre-contact Stó:lō population is estimated to have been between twenty-five thousand and fifty thousand people. I pause to consider such a large group of people in villages along the Chilliwack River.

A network of alpine and riverside trails linked the villages. From Chilliwack Lake, the trails led northwest, down the Chilliwack River to Vedder Crossing, and also south, to the Nooksak and Skagit rivers. The connection in the middle was Chilliwack Lake, crossed in canoes that were cached at each end.

Suddenly, at the end of the eighteenth century, the Ts'elxwéyeqw disappeared from the lake. Two concurrent calamities created the equivalent of a perfect storm for these people. In 1782, a smallpox epidemic spread through the Indigenous communities in the Chilliwack Valley, including the Ts'elxwéyeqw along the Chilliwack River. Two-thirds of the population died of the disease. Around the same time, a rockslide buried one of the Ts'elxwéyeqw villages along the Chilliwack River, killing the inhabitants and disrupting salmon runs. Chief Wileliq moved his people from Chilliwack Lake downstream until, at the end of the eighteenth century, the Ts'elxwéyeqw were located only on the northern reaches of the Chilliwack River close to Vedder Crossing. A Hudson's Bay Company census taken in 1830 shows no Indigenous settlements at Chilliwack Lake and one hundred and fifty to two hundred members of the Ts'elxwéyeqw living at Vedder Crossing. Within a few decades, the Ts'elxwéyeqw were transformed from alpine lake dwellers to a river valley people; from a people connected with the Nooksak in the southeast to one linked to the Stó:lō in the northwest.

The Europeans who arrived in the area in the early 1800s—settlers, farmers, trappers and missionaries—appropriated Indigenous trails as well as the land and language. Ts'elxwéyeqw became Chilliwack. With the establishment of the Hudson's Bay Company trading post at Fort Langley in 1827, the Indian Trail along the Chilliwack River connecting Vedder Crossing to Chilliwack Lake became the Hudson's Bay Company Brigade Trail. Indigenous people carrying fish and roots and berries for trade were replaced by voyageurs ferrying furs to trading posts.

I have a vision of Paul Kane paintings of Indigenous people in my head: the proud warrior sitting on his horse, women and children around a fire in front of a teepee. I want to replace that image with reality, but how to discover the truth?

What did the Ts'elxwéyeqw look like? What did they wear? What did they eat?

Did the lake shape them as it has shaped us?

I pore over the *Stó:lō Coast Salish Historical Atlas* and its maps of trails, historical timelines, charts of seasonal activities. I check out other references: *Landscapes and Social Transformations on the Northwest Coast: Colonial Encounters in the Fraser Valley*, by archaeologist Jeff Oliver; *Among the Ankomenums*, by Reverend Thomas Crosby, an early Methodist missionary in the Chilliwack area; and *Potlatch and Totem*, a recollection by W. M. Halliday, an Indian Agent for the Department of Indian Affairs from 1906 until 1932. I learn that all parts of the western redcedar were used by the Ts'elxwéyeqw: the bark for basketry, mats, nets, clothing; the trunk for bowls, spoons, storage boxes, house posts. Canoes hollowed from forty-foot cedars carried freight and people over the lakes and rivers. Small, shovel-nose

The Ts'elxwéyeqw used nets made of cedar bark attached to the end of a pole to fish in the Chilliwack River. Courtesy of Stó:lō Research and Resource Management Centre.

canoes carved from smaller trees navigated the shallow sloughs of the lower reaches of the Chilliwack River around Vedder Crossing.

The Ts'elxwéyeqw fished year-round, their harvest connected to the migratory patterns of the Pacific salmon. Spawning started on the Chilliwack River in late April. Fish were dried or smoked and stored for the winter months. Kokanee, an inland salmon, was fished in Chilliwack Lake, with spawning starting in early August.

In early summer, salmonberry, thimbleberry, saskatoon berry, trailing blackberry, and gooseberry were picked and dried; later, huckleberry and blueberry. The dried berries were stored in cedar bark baskets to eat through the winter.

When I think back, some of our own traditions were not so different from those of the Ts'elxwéyeqw. Fishing, in particular. Our methods and equipment differed. The Ts'elxwéyeqw dipped nets made of cedar bark attached to the end of a pole into the river.

We cast hooks on nylon fishing lines attached to rods and reels. But for both of us, fishing was essential. My family came to the lake either on horseback or by float plane. We had to fish every day for our next night's supper. Most days, we trolled for Dolly Varden and bull trout in the cold depths of Chilliwack Lake.

From the fishing box, I pull out a wooden line holder, shaped like a fin at both ends, with the green rope line wound around the middle. Attached to the rope is a nylon leader line with sinkers clamped every two feet, and at the end, a shiny spinner hook. In the fishing box are extra line, sinkers and hooks in separate compartments, together with a net and a fishing knife. An open tin sits at my feet beside the fishing box. Earlier this morning, I turned over the compost and slid a few worms into the can. Poppy is sitting in the front of the canoe. He's wearing his navy sweater, zipped up the front against the morning chill.

"Shelley, let's see you put the worm on the hook," says Poppy. "Hold the worm steady and thread it on."

I bait the hook, struggling to ignore the way the worm's guts squish out of the hook holes.

"Good job. Okay, now let the line out."

Gently, I place the hook in the water, letting it trail out behind the boat. I unwind the holder until half the line trails behind us. The spinner blinks like a beacon as it sinks. Soon all I can see is the pale green line drifting behind the boat. I keep my hand on the line, waiting to feel for the tug of a fish. Nothing happens. I gaze drowsily at the water, mesmerized by the ripples fanning out behind the canoe.

Suddenly, I am jolted awake.

"I have something!" I call to Poppy, keeping my voice low.

I pull the line in, curling it in a pile at my feet. The pressure on the line slackens. Has the fish escaped with my bait? Or is it swimming towards me? The fish surfaces with an arcing flash. Pulling steadily, I guide the fish towards the net that Poppy holds at the side of the canoe.

"A good size, Shelley," Poppy says. "One more and we'll have enough for dinner."

I let the line out again as Poppy paddles the canoe along the shore. We troll and wait. There's a lot of waiting in trolling.

I am eight and Poppy is sixty-nine. He is my teacher and taskmaster. If you are going to do a job, do it right, is his motto. He is exacting, but patient too, taking the time to coach me. He insists that I do things right and I am eager to please. I do the same with my grandchildren. I like being the teacher, the storyteller. I make time to take each grandchild out on the lake, or into the forest, to pass down the folklore. I hear myself repeating what Poppy told me.

"Never hand over a knife blade out, always by the handle."

"Watch the wind on the water. If you can see whitecaps, stay close to shore."

"Cup your hand behind a match so the wind won't blow out the flame."

I learned important life lessons from Poppy fly fishing on the Chilliwack River.

In the days leading up to a fly-fishing expedition, we are consumed by endless discussions: forecasts about the weather, debates over what fly the fish are biting. The rods and reels are taken out and inspected. Fly boxes are scrutinized, careful thought given to choosing the flies. Poppy, Mother, Dad, Chris and me. Each of us has our own canvas bag with the essentials: fly box, extra line, pocket knife, net, a couple of peanut butter and jam sandwiches and an apple, tin cup for water. Gran is happy to stay behind in the cabin. She doesn't like fishing.

We motor to the river outlet in Poppy's dark green and black wooden boat with the Johnson five-horse motor. Dad cuts the motor just before the outlet, sets the oars in their locks and rows the rest of the way towards the gravel shore just downriver. Poppy, Mother and Dad, Chris and I clamber out of the boat, pull it up and tie it to the nearest tree. We gather our bags and rods and tramp down the barely-visible trail along the river.

For Poppy, fly fishing is a religion. Every aspect has a set routine. He orders flies with names like March Brown Spider and Griffiths Gnat from Ireland, England and Scotland. They are dry flies that lie on top of the water, as opposed to wet flies that sink beneath the surface. Poppy casts his fly, then twitches it left and right to imitate a fly struggling to raise itself off the water. This will entice a rainbow trout sitting in the deep pools where the river eddies behind a large rock or fallen tree. Poppy has two split-cane rods with cork handles and several Hardy reels. He treats his Hardy reels, made in England, as if they were rare gems. At the end of our expedition, he'll wrap the reels in their purple and gold felt pouches, put them in their boxes and place them on the high shelf in the corner cupboard. The rods will be hung on hooks on the log walls inside the cabin.

We spread out as we travel along the riverbank. The current swirls around fallen logs and boulders creating pools in their lee. Greens and browns intermingle with frothy white. Poppy wears brown, blending in with the river: pants, long-sleeve shirt and cap, leather walking shoes, a vest with pockets for boxes of extra flies, line, a small knife. His canvas fish bag is slung across his back, the fish net tucked in. He carries his fishing rod in one hand, his hand on the reel, facing back so the line doesn't get snagged in the trees.

Poppy stops at a favourite pool, the dark water smooth behind a boulder the size of a Volkswagen that lies two metres off the riverbank. Mother, Chris and Dad keeping going, but I stay with Poppy as he gets ready to cast.

"Approach the pool quietly, Shelley," he says. "Keep low. No quick movements. Make your first cast count."

There is beauty in the fluidity of Poppy's cast. He gathers up slack in the line, the reel zings, his arm sweeps back and forth a couple of times, getting out the right length of line. When the line is ready, he sends the fly out over the river and lands it on the darkest edge of the pool. He keeps the fly in play, hopping it across the water as if it were alive. Suddenly, the line dives beneath the water's surface. The rod bends sharply in an impossible arc. Surely it will break. Holding the rod with his left, and reeling in with his right, he plays the fish in towards him.

Watching him, I hear his voice in my head, "It's a delicate balance. Too much slack and the trout will shake off the fly. Too much tension and the line will break." Even at eleven, I am acutely aware of the skills that my grandfather can teach me.

The trout leaps out of the water, a rainbow of colours catching the sun, then it dives to the bottom. I hope the line holds. Poppy reels in the line in bursts, waiting for the fish to tire. He angles the line so he can scoop the fish into the long-handled net.

"It's a beauty!" Poppy exclaims, the only sound he makes in all that time.

He picks up his wooden cudgel, delivers a sharp blow, and the fish lies still. The trout is about twelve inches long, with a line of rosy red between the gold upper back and creamy belly. Small black spots splatter across its back and radiate down the tail. Poppy stoops to gather moss from the riverbank, and packs it around the fish that he places in the canvas bag. He slings the bag across his back and gives me a smile.

"Now you go off and catch one."

I follow the trail to a log that juts partway into the river where I can get at two deep pools mid-river. I shimmy along the log, and step out onto a rock. Sitting with my back against the log, I pick out a fly from the box and tie it to the line.

I think about Poppy's advice, "No sudden movements."

Slowly, I get to my feet. I do a couple of practice casts in the air, just as Poppy taught me. I swing my arm between eleven and one o'clock letting out the line. At the right length, I let the line soar out across the pool and land the fly in mid-pool. Not perfect. It should have landed near the edge of the pool. I twitch my rod to skip the fly across the surface of the water. I hold my breath. A strike! A streak of colour in the air with a spray of silver droplets. My rod bends sharply downwards. I play the line, alternating reeling it in and letting the fish take the line deep into the pool. Suddenly, the line goes slack. I have lost it.

"The next pool," I say to myself.

We use only a fly on the hook, except in junebug season. Around the third week in July, the junebugs appear. Named for the month when they emerge in most places, junebugs arrive at Chilliwack Lake a month later, probably because we're at six hundred metres. The insect is in fact a scarab beetle, with a hard, dark brown shell. Its wings are less than 2.5 centimetres long, but they rub together to make a distinctive and loud clicking sound. When I hear this sound, I freeze, trying to locate the insect. Carefully I set down my rod, net and bag and sneak towards the branch where the junebug has landed. If I'm lucky, I'll catch it before it flies away. Holding it softly in my hand, wings fluttering against my palm, I make my way over to the rocks where Poppy is standing.

"A junebug," I say, as quietly as I can. He raises his eyebrows and takes out the small tin box with a sliding lid he keeps in his vest pocket specifically for junebugs. He'll put this one on the hook at "the big pool" halfway down the river.

These days, we no longer fish on the lake. We only occasionally fly fish on the Chilliwack River, and never at the river's outlet. Fishing is now prohibited from the outlet down to Slesse Creek, about fifteen kilometres downstream of the lake. The Department of Fisheries built a fish hatchery on the Chilliwack River near Slesse Creek in 1981 to help restore salmon populations. Salmon productivity was being affected by increased activities—fishing, rafting, and kayaking—a sign of the stress on the environment from the press of people. Every few years, my son Michael takes his children far down the river to fly fish. I am not the teacher here; Michael is. He learned from his granddad, just as I learned from mine. I regret that I haven't kept up the tradition of fly fishing. It is not so much the fishing as the time spent on the river that I loved. Poppy died a few years after I returned from teaching in Zambia. My love of fishing was really my love of spending time with him. When he was gone, the fly rod lost its allure.

As I am putting away the *Stó:lō Coast Salish Historical Atlas*, I happen across a map that identifies geographic locations by their traditional names. I run my finger across the map and find Chilliwack Lake with the word Sxotsaqel (pronounced Sko-cha-kel). I look at the key to the meaning of the traditional names. Sxotsaqel means "sacred lake" or "something that is sacred." I sit back in my chair. "Sacred" is what the Ts'elxwéyeqw men, women and children called the lake. That's how I feel about it, too—exactly.

My grandchildren are coming for the weekend. I can't wait to tell them about my research. Just after dinner on Friday night, two sets of headlights shine through the trees.

"Rani, how was the archives?" asks Bronwen, as soon as she is through the door. "What did you find out?"

We sit around the fireplace and I tell them the story of the four bear brothers, with white spots on their chests, who turned into the leaders of the Ts'elxwéyeqw. I tell them how disease and a mountain slide brought devastation to them. How the Ts'elxwéyeqw moved away from the lake down the Chilliwack River valley where they live now.

"So they lived here, just like us?" asks Nora.

"Yes, more than two hundred years ago."

"Wow," says Fiona. "That's amazing."

"You know, my mother and dad used to find arrowheads on the beach near the campground."

Fin looks at me, like he'd like to go out right this minute and start sifting the sand.

Talking about the arrowheads gets me thinking. What happened to these artifacts that Dad found on the beach? The arrowheads were always on the mantle at the Sandpiper. When my parents stopped coming to the lake, the house was stripped bare before Chris and his family took the place over. They did a thorough house cleaning. What happened to the things on the mantle? I must ask Chris if he knows. If we did find them, should we give them to the Coqualeetza Centre?

The next morning, the family gathers on the porch. The branches of the cottonwood shiver in the cool June air, their pale green buds a faint sign of this late spring. A pair of merganser ducks glide past the deadhead. No babies yet. We don't usually see the young until the beginning of July, when the grandchildren have moved in for the summer.

The Douglas squirrels are creating a racket on the porch, scolding each other with a harsh trill. A pair of them runs headlong across the porch and up the nearest fir tree. They stop abruptly, give a trill, then careen farther up the tree, circling until they disappear into the branches far above my head, their fur the colour of bark except for the flash of rust under their tails. Pitted acorns rain down on the roof of the porch.

Fiona laughs watching the squirrels, "They're so noisy."

Her dog, Dennis, darts across the porch, in pointless pursuit.

At the end of the weekend, when Patrick and I drive back to the city, we hear on the radio that the final report of the Truth and Reconciliation Commission of Canada has just been released. Through six years of testimony from survivors, the commission has documented the history of the

Indian residential school system in Canada and has now made recommendations for reconciliation.

"I need to read this report," I say to Patrick. I didn't tell the children about the Coqualeetza residential school, but I tell Patrick now.

The next day I go to my local library in North Vancouver to see if I can get a copy of the final report of the Truth and Reconciliation Commission. They don't have it yet, but they direct me to a recent book about residential schools, *Conversations with a Dead Man: The Legacy of Duncan Campbell Scott*, by Mark Abley. I can hardly believe what Duncan Campbell Scott, Deputy Minister of Indian Affairs in Canada tells a House of Commons committee in 1920:

> I want to get rid of the Indian problem. I do not think as a matter of fact, that the country ought to continuously protect a class of people who are able to stand alone... Our objective is to continue until there is not a single Indian in Canada that has not been absorbed into the body politic and there is no Indian question, and no Indian Department.

I am shocked. Duncan Campbell Scott wrote these words at almost exactly the time Poppy built his cabin at the lake. Do we still feel this way a hundred years later? I don't. But how *do* I feel, and what can I do about it? I go online and order the *Final Report of the Truth and Reconciliation Commission of Canada*, Volume One: *Summary: Honouring the Truth, Reconciling for the Future*.

I know very few people of Indigenous heritage. Their reality is surely different from mine, but I don't know how different or in what way. It reminds me of how I felt when I went to Zambia in 1970. I was an idealistic volunteer, wanting to "do good." But I had no idea how to relate to black Zambians. I had never met a black person, had never had a friend with a skin colour anything but the same as my own. Pale. British. I was an outsider and I felt the burden of Zambia's colonial past resting on my shoulders. Being the only white person in a sea of black was a shocking experience, one that stays with me, one that I am grateful for. I arrived confident that my life was better than theirs, that I knew what was right for my students. I learned that I did not. In those two years, I came to see that Zambian culture had much to teach me. Has much to teach me. I still feel the need to tread lightly when I am in Zambia, not wanting to offend. Not wanting to be the white person who thinks she knows better.

THE SCOUT'S GRAVE

Two grave markers stand side by side, one a cedar cross and the other a marble headstone. Under this photo is a caption: "Commemorating the death of Michael Brown, a scout for the United States Boundary Commission."

I was combing the website of the Chilliwack Archives when I came upon the photograph. It is dated 1936, Chilliwack Lake. How is this possible? My family was at the lake in 1936. Poppy must have known about the gravesite. If Poppy knew, why didn't he tell me?

Clearly, I need to spend some time at the Chilliwack Archives. Located in downtown Chilliwack, it contains extensive research materials and records relating to the history of the area. I add more questions to my notebook.

Who was Michael Brown?

How did he come to drown in the lake?

As soon as Patrick leaves for Vancouver and his week of work, I get in my Prius (Rani's Red Racer, as the grandkids call it) and drive the forty-five kilometres of winding road down to Chilliwack. It is mid-June; the canopy of trees has filled in so that I seem to drive through a sparkling green tunnel. I want to learn about the Boundary Commission. But what I really want to know is whether Michael Brown's marble gravestone still exists.

I arrive at the Chilliwack Archives at nine o'clock. Shannon Bettles, the archivist, buzzes me in and briefs me on archive etiquette: no food or drink, wear the white gloves sitting on the counter when handling materials. I've done my homework online; I hand Shannon my list of sources. She brings a stack of file folders out to me, each one clearly labelled. I lay them out on the heavy oak table, its green leather inlay stained and pitted from years of use for Chilliwack city council meetings.

Within minutes, I find the photo of Michael Brown's gravestone. The inscription reads:

Michael Brown of County Galway, Ireland. CO F 9th Infantry, US Army. Drowned in Chilliwack Lake, October 10, 1858, while serving with his comrades as escort to the International Boundary Commission. His body was recovered and interred here by his comrades. June 18, 1859. Age 26 years.

He has the same first name as my son. I warm to him immediately, this young man dying far from home. I read that he went missing while making a routine crossing of the lake. I sit back in my chair, my imagination kindled. The water frigid, the wind whipping up waves. His canoe capsizes.

I think back to my son, Kevin, travelling in Kenya at the same age. I experienced the anxiety of the unknown the same as Michael Brown's mother. Kevin's routine was to phone us every Sunday. When we last spoke to him, he was planning to cross Lake Victoria on a ferry. Two weeks went by without hearing from him. Then we heard on the radio a news bulletin that a ferry had capsized on Lake Victoria. Many passengers had drowned. I was frantic. Patrick and I spent the weekend at the cabin beset with anxiety. Michael called us. He had heard from Kevin and he was fine. I wept in Patrick's arms. Michael Brown's mother would never know such relief.

I learn a few more personal details from the *US Army Register of Enlistments 1798–1914*. Michael Brown was five feet eleven inches with grey eyes, sandy hair and a ruddy complexion. On March 6, 1857, at twenty-four, he enlisted as a private in the United States Army in Boston. In the summer of 1858, he was one of twenty-five soldiers with the United States 9th Infantry who escorted the survey team to their camp at the south end of Chilliwack Lake.

Ten years earlier, the Oregon Boundary Treaty had established the 49th parallel as the boundary between the United States and the Western Territories of Great Britain. However, the boundary wasn't identified as a physical line on the ground until the discovery of gold on the Fraser River in 1856. In anticipation of the taxes that could be levied on the gold, both countries were keen to locate the border. A Joint Boundary Commission was established.

The American survey team arrived a year before the British and began building an astronomical station at Point Roberts on the Pacific Coast to determine the exact location of the 49th parallel. When the British arrived in the spring of 1858, a disagreement broke out over how, exactly, to mark the boundary. The British wanted to place stone boundary markers every mile. The Americans balked at the cost as well as the physical difficulty of placing markers through the mountains.

As described in Charles Wilson's diary, the boundary markers every mile were eventually determined to be:

> inexpedient at the present time, in consequence of the great expense, consumption of time, and the impractical nature of the country, to mark the whole boundary by cutting a track through dense forest.

A portrait of the British Boundary Commission: standing, Lieutenant Anderson, RE; Colonel Hawkins, RE; Lieutenant Wilson, RE; sitting, Lieutenant Darragh, RE; Mr. Lord, naturalist; Captain Haig, RA: 1859. Image A-05429 courtesy of the Royal BC Museum and Archives.

It was therefore agreed to ascertain landmark points on the line by the determination of astronomical points at convenient intervals on or near the boundary by cutting a track of not less than 20 feet in width on each side for the distance of half mile or more, according to the circumstances.

Instead of a continuous line, the border would be more like a dotted line, with iron markers placed wherever feasible for the western section of the border between Point Roberts at the west and the highest peaks of the Cascade Mountains at the east. Where the terrain was too rough, a mound of stones would suffice.

The survey teams were faced with the challenge of the Cascade Mountains, a volcanic range running parallel to the Pacific Ocean up the coast from northern California into southern British Columbia. A range comparable to the Rocky Mountains further east, the Cascade Mountains were impenetrable from the American side of the border. The only access was from the Canadian side through creek valleys that drained from the Cascade Mountains north into the Chilliwack River.

For the Chilliwack Lake section of the survey, the lake itself was the natural entry point. Both teams decided to establish their camps at the south end. The boundary was only a couple of kilometres further south and adjacent river valleys provided several hundred kilometres of access to the border towards both the east and west.

In the spring of 1858, the American team set up a supply depot at Vedder Crossing where the Chilliwack River meets the Fraser River, 160 kilometres inland from the Pacific coast. Men and supplies arrived by boat up the Fraser, then were transported by packhorse on the Hudson's Bay Brigade Trail along the Chilliwack River to the north end of Chilliwack Lake, arriving not far from where our cabins now stand.

The steep terrain made it impossible to cut a trail to the south end of the lake. The only option was to use the water as a highway. The crews set about building boats and rafts to ferry men and supplies. The American team and the British team each set up camp, one on either side of Depot Creek, which emptied into the southeast corner of the lake.

Hundreds of men lived in these camps. Astronomers, surveyors, topographers, chainmen, instrument carriers, cooks, packers, surgeons, geologists, quartermasters, artists, axemen, labourers and Indigenous guides and messengers. I think of the boggy ground around Depot Creek, thick with trees and mosquitoes and wonder how on earth they managed.

The archivist brings me the diary of Lieutenant Charles Wilson, a member of the British Boundary Commission. Charles was born in Liverpool, England, on March 14, 1836. He always wanted a military career and, at twenty-two, he was accepted into the Royal Engineers. His first overseas posting was as secretary to the British Boundary survey team. In his diary, he recorded not only the progress of the survey but also the exotic natural world he was travelling through, the people and their customs. He kept the diary for his sister back in England to describe his life in the far-flung colony of British Columbia.

> The Chilukweyuk stream is a tributary of the Fraser & rises in the far recesses of the Cascade Mountains & by its valley we penetrate the mountain range later on in the season. The Indians give a wonderful account of a lake near the summit, where bears, marten, marmot & salmon abound; a reconnoitering party sent forward give glowing accounts of the fishing.
>
> The lake is very beautiful in the middle of the Cascade mountains between 3 or 4000 feet above the sea, lofty snowy peaks, glaciers close above our heads, the snow in some places within a few feet.
>
> It was a lovely day, not a breath of wind, the only sounds which

Sir Charles Wilson, age 22, lieutenant with the British Boundary Commission. Wilson's diary provides insight into the natural environment in which the commissions operated. Image A-01625 courtesy of the Royal BC Museum and Archives.

broke the silence being the whistle of the marmot or the rumbling of falling trees or snow in the mountains.

As I was crossing the lake today, one of the sudden mountain storms came on & caught me in the canoe; luckily it blew in the right direction, as nothing could be done but go right before it. I was steering and precious hard work it was; to let the canoe get broadside on in the least would have been an instantaneous upset & the water was so cold there would have been no chance of swimming; besides, we had to work away with the paddles to keep up with the wind or we should have been swamped by the water coming over the stern.

The canoe literally flew through the water with the rain and spray flying past us. It was altogether a most exciting little trip.

Wilson often made the several-day trek back to survey headquarters on the Fraser River, stopping off at camps set up at intervals along the Chilliwack River and named for the officers in charge of the camps: R. W. Haig, captain and astronomer; J. K. Lord, naturalist and veterinary surgeon; and C. J. Darrah, captain and astronomer.

Started off for Lord's camp, 14 miles on horseback & for the first time was brought face to face with master Bruin, which happened in this way; after going over a piece of flat ground formerly the bed of the river & known to us as Grizzly Flat, the trail runs by zigzags up an almost precipitous hill, at the bottom of this I got off to walk up, leading my horse.

The day was excessively hot & I had scrambled down about half way with my head bent down after the usual fashion of getting up steep hills at a quick rate, when I was suddenly jerked into an upright position by my horse starting back, and stood face to face with an old black bear who was quietly walking down & had come upon me round a sudden turn & was now only 6 or 7 feet off. We stood looking at each other for about a minute during which my feelings were anything but pleasant as I had not even a revolver.

If the bear had advanced another step my horse would have backed out of the path & gone rolling down the hill & been killed & my only chance would have been to trust to my heels down the trail; however, Bruin seemed quite as frightened as I could have been & made off along the side of the hill at a pace which astonished me & I pushed on as quick as I could. Though a black bear has seldom or ever been known to attack a man unprovoked, I must say I felt uncommon queer, at such close quarters & was glad when I got to Lord's encampment.

Lord left for Port Douglas & I started back again to spend a few days at Darrah's camp at the mouth of the Slesse river; we have made a bridge over the Chilukweyuk river here, which caused some trouble, we managed it however by felling two trees each about 150 ft long on opposite sides of the river so that their ends rested on a small island in the middle of the river & over this we can now pack mules. Two of our men nearly lost their lives in the operation, having fallen into the rapids from which they were drawn out almost miraculously & had nothing worse than some good bruises & taking in a large supply of water.

Time melts away as I read. Charles Wilson could be describing the lake, the mountains and streams that I know today. I can feel his wonderment and awe. How fortunate am I that I can experience the wilderness much as Wilson did a century and a half before.

The 1858 survey was not the final survey. Before long, the United States raised concerns about errors in the boundary locations in British Columbia. In 1901, Canada and the United States sent a second pair of survey teams to Chilliwack Lake. Once again, they established their camps at the mouth of Depot Creek. The surveyors found the old boundary markers—numbered 50 and 51—on either side of the Little Chilliwack River and replaced them with new markers on the stone piles, numbered 62 and 63. In my notebook, I jot down the new numbers with the question, "Why were the boundary marker numbers changed?"

A third joint survey was undertaken in the summer of 1935. Gordon Watson, a local supplier, was hired by the Canadian team to deliver provisions to the camp that was once again set up at the south end of the lake. In the Boundary Commission files at the archives, I find a report written by historian Neil Smith compiled from interviews with Gordon Watson. Delivering supplies was still a challenge, despite the advances in transportation since 1858. Goods were mustered in Chilliwack, taken by logging railway fifty kilometres, then carried the final six kilometres by packhorse. Amazing! This was exactly the same route my grandparents used. Reading Neil Smith's report brings my grandparents' trek alive.

Why didn't I know about these surveys?

Poppy and Gran must have seen the survey teams paddle by. My mother, too. I imagine the three of them standing on the shore, waving. Why did no one mention this moment in Canadian history? Why was it not recorded in our family guest book? I am frustrated. I could have heard the story first-hand instead of having to do research in the archives. I start to wonder whether perhaps I wasn't paying attention. No, I have always been interested in family folklore, especially about the lake. They simply didn't tell me. I can't believe how easily an event can slip through the cracks of history. Have I left anything untold with my children or grandchildren? Not that I can think of. But I'll keep in mind that I must be sure to tell my grandchildren important events in our history.

The 1935 survey crew was spare compared to the earlier expeditions: only thirteen men, including a cook and cook's helper. Packhorses were ferried down to Depot Creek so that the crew could move camp further south into the valleys on a weekly basis. Gordon Watson built a large raft and floated eight horses, three at a time, down the lake, a crossing of four hours. On the first trip, having set up a sail on the raft of horses, he sailed past the camp at Depot Creek and had to be towed back by canoe with a Johnson Seahorse motor. The crew used an abandoned log cabin at Depot Creek and set up tents in the clearing around the cabin. This must have been Charlie Lindeman's summer cabin, the one he used for trapping marten up Depot Creek.

Gordon Watson moving horses on a raft down Chilliwack Lake for the third boundary survey in 1935. Courtesy of the Chilliwack Museum and Archives, 2011.059.005.

When Gordon Watson arrived here with his horses, he said the cabin and camp were alive with mice and chipmunks. All the grain for the horses had to be put on a raft anchored out from shore.

The men worked an eight-hour day and had to be at the work site by eight in the morning. Sometimes, where they would be working would be an hour or more walk from camp and the same back in the evening. The cook got up early to have breakfast ready for the men. Breakfast would be bacon and eggs, pancakes, jam, tea and coffee. There was a lot of boggy ground near the mouth of Depot Creek, so two bridges had to be built to make it easier for the men and horses to get across the worst places.

The cabin was overrun with mice so Gordon put some water in a 4 gal. can and placed a wire across the top with a baited tin can in the middle that would turn. He would catch 6 or 8 mice a night.

Poppy's mice catcher! I nearly laugh out loud in the silence of the archives. The one we still use at our cabins for catching multiple mice in the night. Poppy must have heard about it from Gordon Watson, or maybe one of the survey crew. Suddenly, I feel a direct and palpable link from me to this third expedition, and all the way down through history to Michael Brown.

The 1935 survey team came across Michael Brown's grave. They reported to the United States government that the cedar cross was badly decayed. The following year, the 9th Infantry shipped a new marble headstone from Washington to Chilliwack, weighing 172 kilograms and inscribed with the same words as the original cross. Charlie Lindeman, the trapper who first guided Poppy to the lake in the early 1920s, transported the headstone by logging railway, packhorse and dugout canoe to Depot Creek and installed it at the gravesite next to the cross.

I am dumbfounded. Charlie and Poppy were close. Poppy *must* have known about the drowning of that young scout and his gravesite. Why didn't Poppy tell me? He passed on lessons about fly fishing but not about dead soldiers. Did he think that a girl wouldn't be interested in such things? My mind is in chaos. So many questions, so few answers.

At noon, I take a break from the archives. I order a sandwich at a nearby deli and sit on a park bench to eat in the sun. I need to stretch my legs, but don't make it halfway down the street before I turn around, too excited to do anything but return to my reading.

I sit down in front of my stack of files, like a squirrel intent on gathering her nuts. I open the file labelled "Michael Brown." I find a hand-written Post-it note stuck to the top paper inside. I am surprised. Tampering with files is frowned upon. The note says, "I was unable to find the headstone and believe it may have been stolen." No name, no date. I slump in my chair, deflated. Could the headstone have disappeared? I go up to the counter and ask Shannon if she knows about the note in the file.

"No," she says. "I'll have to think about what to do with it."

I can't believe that this happened at the lake. In 1935, Poppy was forty-eight, Gran forty-two and Mother, seventeen. Charlie Lindeman was their only neighbour and he came to Cupola Lodge for dinner at least once a week. He would have told them about transporting the headstone. I think of my mother in her nursing home, her brain atrophied by dementia. If only she could remember. There is so much that I want to ask her.

I listen to an archival tape recording of Ray Wells, a rancher from Sardis who lived at the lake in the 1950s and 1960s. He describes Michael Brown's grave:

> Last Sunday I visited a grave on Chilliwack Lake up Brown Crick [Depot Creek] about five minutes' walk from the lake. It's a grave of a young Irishman who belonged to a brigade of American soldiers who were escorting the International Boundary line survey crew. It was their duty to protect the survey crew from the Indians. This young fella got drowned in the lake there. I think it was 1858 he was drowned. They buried him there.
>
> In 1936, the US Boundary line put up a marble slab, nice one, set in concrete, five feet high. It was nicely marked, nicely located. It was back in from the shore in a big grove of cedars. Strange thing about that, too. I've known that grave to be there for thirty/forty years. And it is situated in big timber. And there are windfalls everywhere, all over the ground. But I haven't seen a stick fall across that grave. It's neat and tidy looking, all mossed over.

So, Ray saw the grave. My excitement grows. Maybe it is still there after all. I really want to see this for myself now.

As I listen to Ray's tale, something registers that has been confusing me all week. Many of the place names have two or more names. I've started keeping a list.

Chilliwack Lake, for example, was originally called Ts'elxwéyeqw after the first peoples of the region, the Ts'elxwéyeqw people. In British diaries from the mid-nineteenth century, it was referred to as Summit Lake. By the turn of the twentieth century, Chilliwack, the anglicized version of Ts'elxwéyeqw, was most commonly used. It was officially adopted as Chilliwack

THE SCOUT'S GRAVE ∼ 69

Lake in 1917. Lindeman Lake was originally Silver Lake. Depot Creek became Brown Creek, after Michael Brown, the drowned soldier. The name reverted to Depot Creek after the later survey teams used the location as their supply depot. I knew Paleface Creek as Kokanee Creek when I was a girl and we were catching kokanee salmon on our annual fishing trip down the lake. I don't know when these names changed or why. I add this question to my notebook.

"When and why have the place names changed?"

Before I finish up at the archives, Shannon brings me the International Boundary Commission's *Joint Annual Report 2014*. The boundary of nearly 9,000 kilometres is maintained by the commission on a fifteen-year cycle. The "vista" must be kept open to ensure clear visibility for public notice of the boundary. The commission hires private contractors to cut the trees to maintain a six-metre swath. Where is our section of the boundary in this cycle, I wonder?

I pass my days at the archives, my evenings making sense of the notes I scribble at the leather-topped table. On Friday night, Patrick, both my sons and their families converge on the cabin. Patrick brings the report of the Truth and Reconciliation Commission that I had ordered online and had sent to his office. I'll start reading it later tonight. I am bursting to tell them what I've learned at the archives. As soon as we are all sitting around the fire after dinner, I begin.

"I found out that there was a survey of the border with the United States in 1858," I say. "The survey teams set up camp at Depot Creek near the end of the lake."

"That's interesting, Ma," says Kevin.

"Yeah, they put up stone markers on the boundary line."

"We should go find them," says Michael.

The girls are looking at magazines, checking out their text messages on their iPhones.

"Hey, kids, no iPhones," says Michael. Nora, Bronwen and Fiona put them away and come back to the couch. The boys are playing a game of Hearts.

"I've learned more about the soldier who drowned in the lake," I say a little louder.

The kids sit up in their seats, focusing their eyes on me. Now I've got their attention.

"The soldier's name was Michael Brown. He was from Ireland. He was twenty-six when he drowned."

"What about the grave?" asks Fin.

"I think I know where the grave might be," I say. "A marble headstone was erected eighty years ago. At Depot Creek."

"Let's have a search party," says Bronwen.

"Let's find the grave!" say Roger and Fin in unison.

"Where's Depot Creek?" asks Nora.

The kids head for the computer and look up Depot Creek on Google Maps, to see if the headstone is visible. All they see are trees and marsh.

"You should know," I say, watching this frenzy of activity, "I found a Post-it note at the archives that said the headstone was gone."

No one pauses. Undaunted, they plot how to search the area, where we might start from, how long it might take.

"Maybe we should camp overnight."

"We should bring our headlamps."

"What about our Swiss Army knives?"

"Will we have to bushwhack?"

All of them are talking at once.

"Don't forget the Post-it note that said the headstone was gone," I warn. "Don't get your hopes up. We don't know if it's still there."

But they are so full of optimism and energy that I shed my caution too. If they had their way, we'd set off right away. Me too.

"Let's wait till the end of June, when you're out of school. Maybe your dads can take a few days off work to come too. We can do lots in the meantime."

"Like what?"

"Well, for one thing, I can see if I can find out more about where the grave is located." As I'm talking, it occurs to me that I should ask Kris Sanders. His family has a long history with the Church Camp at the south end of the lake. He might know something.

The last week of June arrives, with wet, cloudy weather interrupted by occasional glimmers of summer heat. The grandkids are out of school now. Kevin, Michael and Molly have taken the week off; Joanna and Patrick, unfortunately, have to stay at work in town.

We're now into the summer routine. Michael's family comes to the lake most weekends and for three one-week stints spread out during the summer. Bronwen, Fin and Will are here occasionally in July and then stay for the month of August. In August, Kevin and Joanna, who are both working, come Thursday night to Monday morning. Patrick is a constant, here every weekend through the summer, taking weeks when he can fit them around his work.

I was able to talk with Kris Sanders on the phone during the week. He said that he was last at the gravesite three years ago and saw the headstone. So the Post-it note writer was wrong. Kris gave me directions to the gravesite. Wow! Having directions makes our chances of finding the gravesite that much greater. I can't wait to tell the kids.

Monday evening we sit around the fire, planning our expedition to find Michael Brown's grave. We'll set off first thing the next day. I start by retelling the story of Michael Brown. The kids are riveted.

"I got directions from the guy who heads up the Church Camp. You remember Kris Sanders?" I say.

"Yeah, I remember him. He came and visited us last summer," says Michael.

"Well, Kris says that we should go by boat to the bay that is immediately south of Depot Creek. Once on the beach, we go into the forest about fifty metres from the shore and look for a flattened space that used to be a forestry campground."

"Where's the grave?" asks Fin. He's impatient for me to get to the important information.

"Just a sec," I say. "Standing in the old campground, we head off into the woods at ten o'clock towards the creek.

I can see confusion in Fin's eyes. He's trying to figure this out.

"The grave is in a grove of large cedar trees."

"What's a grove?" asks Roger.

"A group of trees standing together, separate from the rest of the forest, like maybe in a circle," says Kevin.

"The grove of cedars is halfway between the beach and the creek," I say. "Kris says if we reach the creek, we have gone too far."

As I relay this information, Kevin draws a map, looking up the topographic features on Google Maps. Kevin is the technical expert in the family. Everyone, including GG, turns to him to solve their computer problems.

"Here's where we take the boat in," he says to Michael, turning the screen to him and pointing out the bay to the south of Depot Creek.

"Okay," Michael says. "I'll put extra rope in the boat and fill it with gas before breakfast so we're ready to go first thing."

Michael is the family's organizer and cheerleader. With a flair for poking fun, he teases the kids mercilessly and keeps the energy level high.

"Do you think we'll find the grave?" asks Fiona.

"I don't know," I say, "but we sure will try."

In the morning, we're up with the sun. The girls fill water bottles, one for each person. The boys try on backpacks until they find one that fits.

"I'll set out the fixings for sandwiches," Michael says. "You kids come and make your own."

"I'll put out some granola bars and cookies and cut up some apples," I say. "Everyone put some in your backpack."

I search through my drawers and find bandanas for each of us, in a rainbow of colours and patterns. The kids fashion their best expedition look; the boys tie them around their heads, the girls around necks, arms and ankles. Kevin puts on his *Indiana Jones* hat.

"We look like teams in *Survivor*," says Nora.

Roger and Fin put their Swiss Army knives in a plastic bag in case they fall in the creek. I suggest long pants and shirts.

"We're bushwhacking to the gravesite. There'll probably be mosquitoes."

"Let's take bathing suits," says Nora.

"What about Bandaids?" asks Fiona.

Molly, who is staying behind to study for her accounting exams, starts toasting bagels and frying eggs.

"The kids will need some protein to keep them going."

"Rani, you have toilet paper, right?" says Bronwen.

"Yes, at the front of my pack."

"Okay, everyone, let's get going," says Michael. We each pick up our packs. We have an extra bag with towels and jackets. We troop across the beach and onto the dock where the motorboat is tied up. The dogs have picked up on the hubbub and are dashing back and forth, not wanting to be left behind. We pile into the boat, Michael at the helm, Kevin casting off. We wave goodbye to Molly.

"We'll aim to be back by five," Michael calls out to Molly.

"Have fun! Good luck!"

I can feel the buzz of excitement in the boat. The dogs pace from bow to the stern, finally curling up and putting their heads down. The day is overcast with a slight wind and chop on the water. Michael pushes the throttle down and heads straight down the middle of the long, slender tongue of water. The kids huddle together to keep warm.

We pass the mouth of Depot Creek and motor into the bay. Michael cuts the throttle and Kevin points out a good spot to bring in the boat. Michael steers slowly towards shore, as Kevin keeps an eye out for submerged logs. Michael steers the boat onto the beach. The kids and dogs clamber out and dash into the forest in all directions.

"Hold on kids," shouts Michael. "Let's get organized. Mom, what were the directions? Can you tell us again?"

The kids assemble in a troop. I read out the directions. We set off in a more orderly fashion towards the campground circle. The forest immediately becomes impenetrable, with fallen trees covered by a thick layer of lichen and a bog underfoot. We can't find anything that resembles a campground circle. We walk single file along a fallen tree about a metre off the forest floor, above the giant floppy leaves of skunk cabbages and masses of huge maple-like leaves with spiky thorns and bright red berries clustered at the stems.

"Oh no," says Fiona. "Is that Devil's Club?"

"Yes," Kevin says. "Just don't touch the leaves or stem."

A scratch of Devil's Club can produce a nasty sting and rash worse than stinging nettle.

"I'm not happy about this," says Fiona. Fiona is like me, cautious, wanting to be sure before proceeding. I push myself to be daring. I tend to push Fiona, too. But Fiona knows her own mind. She won't let me push her where she doesn't want to go.

We keep in sight of one another, Michael leading one group of kids along a fallen tree route and Kevin along another. After about a half hour, we stop and look around us. We faintly hear the gurgle of a creek.

Michael pauses. "I think we're in the right place. I see cedar trees ahead. I don't think they'd have dug a grave in this swampy area. There has to be higher ground ahead."

The creek gets louder. We hear a whistle. Through a thicket of branches, we see Kevin standing on a fallen tree ahead. He is waving.

"Over here!"

We rush forward along another fallen tree, trying to avoid the Devil's Club and keep our footing. And then it comes into view. A marble headstone, shimmering white amidst the green and brown of the soaring cedars.

We scramble down and crowd around like acolytes. The kids are looking at me with shining eyes. We have found the gravesite. I feel like the Pied Piper. The kids and the boys followed me here. This is exactly what I dreamed of: an adventure that none of us will ever forget, that connects us to the history of the lake.

They all talk at once.

"I can't believe we found it."

"Let's read what it says."

"Are there bones under my feet?"

Michael Brown's cedar cross disintegrated and was replaced by a marble headstone in 1936. The author and her family find the headstone buried deep in the forest at the south end of Chilliwack Lake. Courtesy of O'Callaghan family.

The stone not only exists, it is pristine, no discolouration, no dullness to the crisp words chiselled into its surface. I read out the inscription. A young man from County Galway was buried here one hundred and fifty-five years ago.

I want to stand here forever to etch this moment into my brain. The grave is more than just a symbol of the past. Of the pull of adventure, the desire to know, to see, the willingness to take risks. It speaks of curiosity, perseverance, resourcefulness, traits that I hope these children will carry into adulthood. On the other hand, it is a reminder of a time when boundaries were drawn, when Europeans, whose own land had run out, came and took the land from the people who lived here first. Michael Brown was a scout, his job was to protect the surveyors from the Indigenous people. I think back on my research at the Coqualeetza Centre. We commemorate the Irish scout. What about the Indigenous people who died here? Where are their headstones?

"This is awesome!" Bronwen says, bringing me out of my reverie.

"It's like our own special place," says Nora.

The others nod their heads, looking around them.

As we walk back, I tread lightly. "Let's spread out a bit," I say. "We don't want to make an obvious trail." Maybe the monument is still here after all these years because it is remote, hidden away in the forest. I want to leave it that way, so my grandchildren can bring their grandchildren here too.

When we arrive at the boat, I pause to look back into the forest. The black trunks of the trees create silhouettes against the sun's rays filtering through the forest. It feels mystical, sacred. I can hardly believe we found the headstone.

Kevin is waiting to help me into the boat. "That was amazing, Ma," he says. He unties the rope, tosses it on the seat and climbs in. As I climb in, I suddenly realize something.

"Kevo, we forgot to look for the wooden cross."

"You're right," he says. "A good reason to come back."

We were so overwhelmed by finding the headstone that we forgot all about the cross. I wonder if we could have found any remnants?

The clouds have darkened and the wind is whipping up whitecaps on the water. Shafts of light stab through the clouds. We are now on the search for the markers showing the boundary between Canada and the United States. Michael starts up the motor. In a few minutes, we are at the southernmost end of the lake. Michael pulls into the middle of the sandbar beach. He lets us out, then comes to the bow of the boat to talk with Kevin and me.

"We can't leave the boat here. The waves will push the stern around and the boat will be pounded."

He's right. Waves are rolling steadily onto the beach. Wind whips the girls' hair about their heads.

"Should we abandon the search and head for home?" I ask.

Michael and Kevin huddle.

"No, I don't think so." Michael says. "Why don't we tether the boat to that deadhead?" He points to one about a hundred metres from shore.

"Yeah, that could work," says Kevin.

"Kev, you're good at knots," says Michael. "I think you should go."

"You're right. Okay."

Kevin strips down, pulls on his bathing suit and drives the boat out to the deadhead. Into the water he goes with the ropes and secures the boat to the deadhead. When he climbs back in, he stares, shoulders set. The waves are quite high now and the boat is jerking sharply against its mooring. Michael shouts at him, but the wind is blowing too hard for Kevin to hear.

"I better go out and help him," Michael says. "Something isn't right."

Michael swims out into the choppy water. The two of them work together in the water, diving down to secure the tether rope. After about thirty minutes, they swim to shore.

"It'll be good," Kevin says. "I was worrying about one of the limbs on the deadhead under the water, but we positioned the rope so the boat is secure."

"Man, that water's cold," says Michael, towelling some warmth into his skin.

"Good job, boys," I say, confident in their judgment call on the boat. We put on our packs and head out. I am relieved we don't have to abandon our expedition. I'm as anxious as the kids to find the boundary markers.

With the help of Kevin's GPS, we find the trail to the border. Fiona notices a sign: *Chilliwack River Ecological Reserve.* The trail is marked with small red squares on trees at eye-level and orange marking tape. We walk through a forest of cedars, some as thick as ten metres in diameter at the base, the trunks towering high into the sky. The limbs are covered in hanging lichen that Poppy called "Old Man's Beard." We are surrounded by a bewitching forest, with the huge arms of the trees dripping tangles of pale yellow-green. The trail winds its way closer to the river where the brush would be impenetrable if not for the cleared path. At times, the alder thickets are so close that I can't see more than a metre ahead.

We scramble down the bank to the river to find a spot for lunch. We perch on a log fallen across the river and catch the sparkle of rushing water beneath us. The dogs bound into the river and lap loudly.

"How far is the border?" Will asks. "When will we get there?"

"I think we're pretty close," Kevin says. "Maybe another half hour?"

"Will we see a sign?"

"I doubt it, not out here in the wilderness."

We eat our sandwiches in a line along the log.

"Are we nearly ready to go?" I ask.

"Not quite, Rani," says Nora. "Who's got the Oreos?"

"They're in my pack," says Michael. "Here, pass them around."

The sun comes out and the leaves glitter against the bright water.

After lunch, we don't have to walk far before we stumble upon a nine-metre swath cleared through the forest. The 49th parallel: the wide line that defines where our country ends and another begins. The cut swath has filled in with new growth but the borderline is still clear. Maybe we're halfway through the fifteen-year cycle? I wonder whether Poppy purposely chose a summer home so close to the border with the United States. But we have no family, no connections at all south of the border. We are staunch and proud Canadians. No, his choice was happenstance.

"So I walk across here and I am in the United States?" asks Nora.

"Yes." I say.

"I feel strange," says Bronwen. "It's so open, there's no one to stop me."

"Yeah, what if we were bad guys smuggling drugs," says Roger.

"You'd have a long way to haul your stash," says Michael.

"Don't even think about it, Rog," Kevin says.

"Imagine if the border were fifteen kilometres further north," I say. "Our land would be in the United States."

"That'd be weird," says Fiona. "I like that we're in Canada."

"The boundary markers should be right here beside the trail," I say.

"Well, we can't see them," says Kevin. "Let's go off the trail a bit."

We bushwhack off the trail to look for the boundary markers, which the survey file described as cairns made of large stones with a bronze marker set in concrete on the top. Kris Sanders had said that he could see them from the trail when he hiked in here ten years ago. We tramp to the edge of the river. Kevin and some of the kids cross the river to see if they can find anything on the other side. The journals in the archives describe a 1.5-metre stone marker on each side of the river. They should be obvious, but we can't find them.

"Should we retrace our steps?" says Kevin. "Walk along the boundary line again?"

"I don't know," I say. "It's getting late."

"Yeah, it'll give us a good reason to come back," says Michael.

I don't push to keep looking. For me, the boundary markers are secondary to today's expedition. My primary goal was to find the scout's headstone. And we did. We can come back another day for the markers.

"Let's walk back through the river," Fin says.

The other kids take up the idea. Kevin and Michael look at me. I smile and shrug my shoulders. Off we go, plunging into the water, soaking our shoes, pants, shirts. At first, we wince with the cold. We scamper onto a gravel shoal on the river's edge, hopping up and down with pins and needles in our feet. We plunge back in. After ten minutes or so, we don't even notice the cold. We're used to cold water at the lake. We're not sissies, we're tough. I imagine that for some, this would be extreme. Am I being a drill sergeant? "Get into that creek, kids! What's a little cold water to us O'Callaghans?"

The river is shallow, with a mild current and occasional deeper pools. In the rocky shallows the water is topaz; in pools, jade green; at the sandy bottom, pale emerald. Sunlight plays on the water, first bright and full of colour, then muted behind a cloud. Rain showers us briefly, but not enough to dampen our spirits. The kids splash down the river with abandon, not caring whether they are up to their knees or their waists.

"Quick! Someone grab Will. He's losing his footing," Fiona says.

Nora grabs the back of his sweatshirt. Will seems unperturbed by being swept along.

"Scooch," Nora says, "stay on your feet."

Michael hears a rustling ahead. Swiftly, he pulls us together. We keep walking, but cautiously. Around the corner, we find a half-eaten fish and fresh bear prints in the sand.

"Are we okay?" says Fiona. "Will the bear attack us?"

"No." says Michael. "Make lots of noise. And stick together. A bear will stay away from us."

Michael leads the group, Kevin bringing up the rear. We start whistling and singing, nervously scanning the bank. We spot more bear prints along the route, but no bears. We gain confidence as we carry on. Even so, I'm happy to see the trailhead at the end of the river.

"The boat's still here!" shouts Roger.

Another reason to be thankful.

Kevin dives back into the water to untie the boat. The kids can't believe how warm the water is after the cold river. They throw off their shirts and pants and jump in, washing off their muddy legs and shoes. Then we load into the boat, wrapping ourselves in towels and extra coats as we motor back towards our cabin at the north end of the lake. As we motor close to Depot Creek, I stare into the forest, imaging Michael Brown's pearly headstone.

"The grave's in there," Fiona says, pointing to Depot Creek.

"Let's go again," says Fin.

"And look for the wooden cross, eh, Ma?" says Kevin.

"We'll go back again, for sure," I say.

As we pull into our dock, the kids start shouting to Molly, who has seen us coming and has walked down to the dock.

"Mom, it was a blast."

"We found the grave."

"We walked down the river."

"We nearly saw a bear."

Molly smiles at her kids' excitement and looks up at me.

"Finding the headstone was amazing," I say. "Pristine white in the middle of the forest." I'm thrilled with the day's expedition. As I walk across the beach to the cabin, the old dugout catches my eye—the next puzzle to be solved.

CHIEF SEPASS

July first, Canada Day. Patrick and I take our coffees to the end of the dock. The sun's rays creep along the shore and capture us at precisely eight o'clock. I turn my face to the warmth and let it seep into me. The water is calm, not a sound except the lapping of the water against the dock. Our small Canada flags wait in our laps. One by one, the rest of the family emerges from the cabin. They are dressed in red and white T-shirts and bathing suits; each of them carries a flag. When everyone is gathered at the end of the dock, we stand and face the lake and sing our national anthem. It seems strange, now that I think about it, facing the United States. But we have always faced the mountains to the south, drawn instinctively to their majesty. Kevin, as always, sings with gusto. His fourteen-year-old niece, Nora, raises her eyebrows, turning to Bronwen, his thirteen-year-old daughter, who shrugs her shoulders. Patrick's mum, called GG by her great-grandchildren, is here with us for the weekend as she often is in the summer, even now at ninety-one. Fin holds GG's arm to keep her steady on the dock. She calls today Dominion Day. Although the name was changed to Canada Day in 1982, GG prefers the original.

"Happy Dominion Day!" says GG.

"Why Dominion Day?" asks Fin, still holding GG's arm.

"We were the Dominion of Canada when our country was first created," says GG. "The Queen of England was our queen, too. I like that."

After we sing, Patrick lines us up for a photo.

Oh Canada. We are proud to be Canadian: tolerant, peaceful, embracing the cultures of others. After the family photo, the others disperse in search of breakfast, but I linger on the dock staring down the lake. Tolerant? Peaceful? Embracing? I'm deep into the history chapter of the *Final Report of the Truth and Reconciliation Commission of Canada* and I don't think our relationship with Indigenous people is anything to be proud of. I certainly don't want our relationship to be defined by the residential school system. But I'm flummoxed as to how to reconcile my pride in being Canadian with this history. I guess acknowledgement is the first step. But what is the second step?

The breakfast gong disturbs my reflections. I look up and see Will standing on a bench on the porch stretching up to reach the gong. He bangs away with all his might. I give him a wave.

"I'm coming."

I step off the dock, cross the beach and climb the steps to the porch. In the kitchen, Michael fries bacon on the wood stove. Molly scrambles eggs and toasts English muffins. Will was a little premature on the breakfast gong, I see. We have breakfast on the porch in the sunshine. As always, a squabble ensues among the kids as to who should do the dishes. Fiona and Bronwen concede that it's their turn. After breakfast, we tackle jobs for an hour, a lake tradition. Fin, Roger and Will fill the wood boxes and cut kindling. Will prides himself in stacking the wood just so.

"Great job, Will. You are definitely the best wood stacker in the family," says Patrick.

Roger jumps into the wheelbarrow and Fin rides him around in circles, dumping him on the ground.

I meet the kids at the shed, pass out rakes and we set off to clean the paths of windfall. We have several paths: to the outhouse, the boathouse, the tent, the water intake up Cupola Creek and a longer one to Otter Cove. We start with the Otter Cove path. Roger pitches pine cones; Will and Fin join in the pinecone war.

"Rani, send the boys ahead. They're a pain," says Nora.

Fiona and Bronwen lag behind, hoping I won't notice. Eventually, though, everyone gets down to work raking debris off the path.

"Doesn't the path look better?" I ask when we arrive at Otter Cove a half hour later.

I like things neat and tidy but I'm not sure the kids share my enthusiasm. My desire for orderliness reminds me of Poppy. Once I take on a task, I am driven to complete it, and not only complete it, but perfectly. I am not easily distracted. On my mission this summer, this trait is helping to propel me forward. But it can also make me be rigid. I should be the student as well as the teacher. I should learn from my grandchildren to be flexible.

At Otter Cove, they toss their rakes to the ground and clamber down to the shore. I follow. The kids take off their shoes and socks, roll up their pants and wade into the water. The boys skip stones. Fin has the best arm: he easily holds onto his record of twelve skips.

I perch on a log and watch the kids play, the lake a backdrop to their antics. My head is filled with questions. First and foremost, the canoe.

Who built our dugout?

How did Poppy get it?

The next morning I get up early. The light is pale, with the sun not yet over the ridge. Recent rains have overwhelmed the creeks. I can hear the rushing and tumbling of Cupola Creek coming down the hill behind the cabin, spilling its torrent of frothing water into the lake, spreading concentric rings of white bubbles. A family of twelve merganser ducks scoots across the water, searching for minnows, quite a brood for the mother to shepherd. The mother clucks as she spies me walking down to the beach, and the babies speed to her side, the smallest one jumping onto her back, the rest in a tight line behind her.

After a quick breakfast, I say goodbye to Patrick and drive down the road in the shadow of overhanging branches, the new leaves a brilliant emerald. I have decided to go back to the Stó:lō Research and Resource Management Centre, this time in search of the spirit of Chief Sepass and his connection to our dugout canoe.

I can hear Poppy's voice in my head, words that play like a repeating record as I drive into town: *Indian Billy carved our canoe and gave it to me in payment for a debt.*

That is all my memory will dredge up. Nothing useful, like when the canoe was carved, or when Chief Sepass gave it to Poppy, or the reason behind the debt. Is Indian Billy the same person as Chief William Sepass, or am I forging links where none exist? Is the story even true, about the building of the canoe? If Chief Sepass did carve our dugout, does it rightfully belong with the Skowkale First Nation? Should I return it? I've considered contacting the descendants of Chief Sepass, but I don't know how to go about finding them. And if I did, what would I say?

Tia Halstad, the librarian at the Stó:lō Centre, welcomes me back. I tell her about my search for information on Chief Sepass.

"The one who wrote *Sepass Poems*," she says. She leaves and returns a few moments later. "We don't have any books specifically on Chief Sepass, but these books on the early history of the area may have some references."

She brings me *The Chilliwacks and their Neighbours* by local historian Oliver Wells and *The Chilliwack Story*, edited by Ron Denman, director of the Chilliwack Museum and Archives. I dive in, hunting for answers, but I find nothing beyond the basic facts of Chief Sepass's life.

His full name was William Sepass Skulkayce of the Skowkale First Nation. He was born in Kettle Falls, Washington, in 1841 and lived for more than a century. His mother was the daughter of a chief of the Thompson River people (now called the Nlaka'pamux Nation); his father, the son of a Shuswap (now called Secwepemc) chief, one of the Indigenous peoples of the interior of British Columbia. He was destined to be a chief coming from such a lineage. Chief Khal-agh-thit-til of the Skowkale was William

Chief William Sepass became both the elected and hereditary chief of the Skowkale First Nation at age twenty-five. His first wife, shown here, died at a young age, along with twenty-one children, of tuberculosis—a disease introduced by early white settlers in the Chilliwack area. Courtesy of the Chilliwack Museum and Archives, AM 362, File 293.

Sepass's uncle. According to Skowkale tradition, a chief chooses his successor, considering family heritage as well as intellectual and spiritual promise. Chief Khal-agh-thit-til saw these qualities in his nephew. When William was just a boy, his uncle began to pass on traditional oral narratives and songs to prepare him to be a leader of his people. At twenty-five, William Sepass was elected chief of the Skowkale First Nation. At the same time, he became a hereditary chief of the Ts'elxwéyeqw people.

What is the difference between an elected chief and a hereditary chief? *The Chilliwack Story* helped me understand. The role of hereditary chief was traditionally passed from one generation to another. The Indian Act, enacted in 1876, required a chief to be elected every two years. Hereditary chiefs were set aside. The elected chief, in conjunction with the band council, was responsible for administration of band affairs, such as schools, housing, water, sanitation and roads. Many bands, however, have carried on the concept of a hereditary chief who provides spiritual as well as secular leadership.

Chief Sepass was given the hereditary name K'HHalserten, meaning "Golden Snake." K'Hhals—Sun God—is the central figure in the traditional narratives of the Skowkale and is part of a hereditary chief's name.

Is Chief Sepass the same "Indian Billy" that features in Poppy's story? He was forty-six years older than Poppy. When Poppy arrived at Chilliwack Lake, Chief Sepass was in his eighties. Could he have carved a canoe at that age? It seems unlikely. Why would Poppy have called him "Indian Billy" when he was so much older than Poppy. And he had been a chief of his people for over fifty years. I just don't know. I look for other chiefs named William or Billy in the Chilliwack area, but find none.

I am uncertain how to refer to Chief Sepass. He was called "Indian Billy" by Poppy and I find other references to "Indian Billy" in reference to William Sepass in historical writings. I find this offensive, treating a man as a child. The name reminds me of white colonials in Africa referring to black men—workers, students, farmhands—as "boy." I encountered and railed against this same practice when I taught in Zambia in the 1970s. Poppy and his generation's relationship to Indigenous people reflected an innate racism, just as was displayed by Europeans in Africa. But that is the past, and I want to be respectful. I will call him Chief Sepass.

In the archives, I study a black-and-white photo of Chief Sepass. No date is provided, but he looks to be in his mid-twenties, around the time when he would have become chief. I can see in the head-and-shoulder shot that he is wearing European clothes: a white shirt and black bow tie, a tightly buttoned black jacket. He is small in stature, just over 1.5 metres, but his erect carriage signals strength, reinforced by the direct gaze of his deep brown eyes. His mouth is resolute under a thick, bushy mustache that hangs like a shelf over his wide mouth. His shoulders are squarely set and well muscled, like those of a boxer. A young man of seriousness and substance.

While I search through the BC Archives database, I come across a series of interviews recorded by Imbert Orchard in 1964, part of oral history interviews conducted for documentary radio programs for the CBC in Vancouver. I send away for a tape with an interview of Ray Wells. In this tape, Ray Wells describes Chief Sepass:

> Old Chief Billy Sepass was one of the Indians I knew the best, and he worked for my grandfather and my father too. We used to go on lots of huntin' trips together. Well, he was a stocky built man. Ah, not what you'd call a powerful man. Stocky built but medium strength. I would say he was very intelligent, highly respected by all Indians and all Whites. He was a very influential Indian, and widely known up and down the coast.

Chief Sepass experienced first-hand the devastation wrought by the diseases introduced by white settlers to the Chilliwack Valley: tuberculosis, smallpox, measles, influenza, whooping cough. A devastation that began in the early 1800s and continued well into the twentieth century.

Ray's tape continues:

> He had three wives, and, he had twenty-four children and I think, ah, there's only two or three alive today. Tuberculosis was runnin' pretty strong amongst the Indians. It cleaned out most of Billy's family.

Chief Sepass experienced epic changes in his lifetime. The traditional life of his people was annihilated. Everything they knew to be true was turned upside down. From 1850 to 1900, Chief Sepass witnessed the arrival of the first settlers and the Christian missionaries, the influx of prospectors fuelled by gold fever, the creation of reserve lands for the use of Indian bands and the establishment of Indian residential schools. From the time of Confederation until the Second World War, Chief Sepass had the heavy responsibility of leading his people through the morass.

Chief Sepass converted at an early age to Christianity and became a leader in the Wesleyan Methodist Church in Sardis, a small white settlement in the traditional territory of the Skowkale First Nation. Methodist missionaries arrived in Sardis in the late 1860s, intent on converting the Indigenous population to Christianity. In 1884, the Methodists opened the Coqualeetza Industrial Institute in Sardis, now part of the Coqualeetza Cultural Education Centre complex in which I am sitting. At the archives, I uncover conflicting stories about how the school acquired its land. In one version, Chief Sepass gave the land to the Methodist missionaries. In another, the Methodists bought the land from the government. Some of Chief Sepass's children may have attended the school. I find a photo in the archives showing Chief Sepass sitting on the steps of the school with his daughter and the principal. What, I wonder, would have led Chief Sepass to support a school where children were separated from their families and forced to trade their language and traditions for the white man's culture? I read that the Coqualeetza School was considered one of the most enlightened schools of its kind. In Chief Sepass's eyes, would the benefit of education outweigh the harm?

Settlers arriving in British Columbia had a voracious appetite for land. By Royal Proclamation of 1763, all land in the Americas was claimed by the British Crown to be held in trust for Indigenous peoples. After British Columbia became a colony in 1858, Indian reserves were created but in quick succession were reduced in size. Other provinces of Canada allotted families on reserves between 160 and 640 acres per family. In British Columbia, this was reduced to 20 acres per family. The view espoused by government officials at the time was that white settlers would make more productive use of this land. From the late 1860s onward, Indigenous leaders protested the size and location of their reserves. The McKenna-McBride Royal Commission on Indian Affairs in British Columbia was established in 1912 to resolve the "Indian reserve question." Hearings were held over three years throughout the province.

Chief Sepass made a presentation before the commission. He felt that the bargain he'd made on behalf of the Ts'elxwéyeqw people had been betrayed.

> Sir James Douglas was the one that surveyed this property for us. The grievances which I am laying before you is what I have already said. After this reserve was surveyed for me by Sir James Douglas, from then I came to learn that there would be compensation made to us Indians for all the land in the Province. I made my complaint to Sir James Douglas and I wish these lands to be returned back to me.

His people's land had been stolen. During Chief Sepass's lifetime, from the 1840s to the 1940s, the Skowkale moved onto two reserves in Vedder Crossing. Chief Sepass's pleas fell on deaf ears. Although the commission nearly doubled the area of reserve lands in British Columbia, the value of the land added was only a third of the value of the land taken away. No compensation was given for what was lost. And what was lost was everything to these Indigenous peoples whose livelihood and spiritual health derived from their connection to the land. In later years, the number of reserves for the Skowkale First Nation was doubled to four, with a total of fifty-eight hectares, where the Skowkale First Nation continue to live today. The traditional territory of the Ts'elxwéyeqw, of which the Skowkale First Nation forms a part, covers 95,000 hectares extending up the Chilliwack River and surrounding Chilliwack Lake. A population in the twenties of thousands in the 1800s is reduced now to five hundred and seventy-one registered members in 2011.

The two cultures had a diametrically opposed perception of land ownership. Private ownership and territorial delineation was valued by white settlers. Shared ownership of commonly held land was valued by Indigenous peoples. As a lawyer, I studied land claims cases. Intellectually, I understood the course of history and the resulting creation of reserves. But sitting in the reading room of the Stó:lō Centre, I am stunned. It is all so wrong. We were a wrecking ball destroying Indigenous traditions and culture. We stole their land. I don't know what to do with this realization. Is this a cop-out, that I throw my hands in the air? I have no answers. Should I have answers?

Throughout his life, Chief Sepass carried on the traditional pursuits of fishing and hunting along the Chilliwack River and at Chilliwack Lake. As a young man, he worked at the Wells's family farm in Sardis, learning the latest techniques for running a modern dairy farm. In his forties, he established his own farm on the Skowkale Indian Reserve in Sardis. He built a house surrounded by hedges, a fruit orchard and a barn, and tilled his fields.

The librarian appears at my side, brimming with excitement.

"Here's something that could be interesting," she says. "A paper on the Sepass canoes."

I eagerly open the file folder. Inside is a manuscript: "Bring Home the Canoe: History and Interpretation of Sepass Canoes in S'ohl Temexw," by Madeline Knickerbocker, PhD candidate in the Department of History, Simon Fraser University, published in 2011.

The paper explores the history of a canoe carved by Chief Sepass, created in 1915 and used by Chief Sepass for fishing and transportation. The author interviewed Bill Sepass, who said that his great-grandfather, Chief Sepass, considered the canoe his prize possession. On Chief Sepass's death in 1943, the canoe was acquired by A. C. Wells, Ray Wells' grandfather. There is uncertainty as to whether the canoe was bought, as is contended by the Wells family, or taken, as is intimated by the Sepass family. In 1962, the canoe was donated by Oliver Wells, Ray's brother, to the Chilliwack Museum and was put on display. In the 1990s, the canoe was moved to the Xa:ytem Longhouse and Interpretive Centre in Mission. When the Xa:ytem Centre closed in 2010, the Sepass family petitioned to have the chief's canoe returned to the Stó:lō Research and Resource Management Centre where it is now housed in the archives building.

I take a break at mid-day and go to see Chief Sepass's canoe. Resting on v-shaped wooden stands, it sits low to the floor so you can see inside. The showpiece of the educational centre, it is lit from above, with descriptive signs designed for tours of school children. Seven metres long, one metre wide, and eight centimetres thick, the shovel-shaped canoe is a burnished brown cedar and finely carved. It has two cross-seats mid-canoe and a centre cross-beam with a hole in the middle for a sail.

I am struck by the differences between this canoe and our dugout. The canoe at the centre is a racehorse, with fine lines, artistry and beauty, compared to our workhorse at the lake, plodding and plain. Seeing this masterpiece makes me question Poppy's story. How could our rough-hewn boat have been wrought by the hands of the master carver, Chief Sepass? I am beginning to doubt that Poppy's "Indian Billy" was the same person as Chief Sepass.

I must have paddled in our dugout close to a thousand times with Poppy. Carved out of a single cedar log split in half, it is four-and-a-half-metres long, a half-metre wide and five centimetres thick, with a rugged board seat at each end and open in the middle. The adze cuts have created a rough surface like ripples on a pond. The cedar is charcoal black, burnt to a depth of half an inch. If I catch my fingernail and a splinter dislodges, I can see the pale yellow cedar underneath. The canoe sits low in the water so that, when paddling, I'm part of the lake itself.

At the beginning of every summer, Poppy inspects the canoe and applies a coat of pitch over the hull. The canoe is set upside down across two saw-horses on the beach. Poppy has a small fire burning. Balanced on a metal grate over the fire is a can of coal tar. When the tar is hot and the consistency of thick paint, Poppy takes the can off the fire and sets it down in the sand. He puts his brush into the tar and slides it across the bottom of the canoe, until it looks as if the canoe bottom is a Dilly Bar dipped in chocolate.

"We need to seal the bottom," Poppy says to me, "or the canoe will leak."

One summer, he finds a narrow split in the hull. Every time Poppy and I go out for a paddle, water seeps up over my toes. I bring along a tin can for bailing. Finally, Poppy hauls the canoe out and stuffs strips of sisal matting in the crack and applies the tar over top to create a smooth shiny black surface. He adds the coating of tar onto our wooden rowboat as well. The sharp smell of the tar reminds me of road repairs in the heat of summer.

Beside the canoe at the centre is a plaque describing the carving process of this lovely specimen. When he was seventy-four, Chief Sepass felled and carved a redcedar on the banks of the Chilliwack River.

I imagine Chief Sepass planing the bottom of the tree, creating a smooth surface with a slight rise towards the bow and a lesser taper towards the stern. With the canoe upside down, he rubs the surface smooth with the flat side of a rock, rough like a rasp. Once the bottom is chiselled, he turns the canoe right side up. He spreads cedar chips along the full length of the canoe and lights them. He allows the fire to burn just slightly into the wood, then removes the chips. He presses vine maple sticks young and green, cut to a length slightly wider than the canoe, into a half moon position. When the shape has spread sufficiently, he allows the fire to die out. He leaves the canoe to cool in a cold rain, with the maple sticks acting as thwarts to preserve the shape of the canoe. He paints the inside of the canoe brick red, using an iron pigment paint made with the natural rust deposits along the Chilliwack River. The whole process takes most of the summer. When the canoe is finished, Chief Sepass paddles it downriver and out into the Fraser River to tend his fishing nets.

Back at the cabin at the end of the day after my visit to the Coqualeetza Centre, I go out onto the beach and look at our dugout. Its imperfections jump out at me. I am not convinced that Chief Sepass carved our dugout. If he did, it was done in a hurry without much care, something I prefer not to imagine. Our canoe is crude and unpolished—not worthy of the master carver.

The next morning, I walk over to the Sandpiper to look at the guest books that Mother and Dad kept. I flip through looking for entries that mention the canoe. I find one in my mother's handwriting in August 1977:

> I met a relative of Chief Billy Sepass who made the dugout canoe. She said there was a publication of his work called *Sepass Poems* and the Coqualeetza Centre is doing a history of this area.

So maybe our dugout was carved by Chief Sepass. Or at least my mother thought so. This is becoming more and more labyrinthine.

In 1916, at seventy-six, Chief Sepass was still a vibrant leader, deeply involved in the welfare of his people. He looked for ways to bridge the gap between his people's poverty and the wealth that the white man seemed to conjure from the wilderness. Legends of gold at the south end of Chilliwack Lake had been passed down through generations of the Ts'elxwéyeqw. In the middle of the First World War, Chief Sepass persuaded a group of local miners to trek to the mountains near the border in search of minerals. One of his partners was Reverend George Wright, who would return later and start the Church Camp at the lake. They did not find gold, but there were traces of silver, enough to support a small camp of prospectors who lived beside the Little Chilliwack River.

In his interview with Madeline Knickerbocker, Bill Sepass talks about other canoes that his great-grandfather carved:

> There was a freightliner canoe that they used which old Chief Sepass had made where he had found a silver mine at the other end of Chilliwack Lake. He named it the Silver Chief Mine. I don't know what business ventures he had with the miners at the time, but he used the freightliner canoe. He made a twenty-one-foot long cedar canoe to transport the people and goods across, back and forth, on the lake.

In the second weekend in July, despite a drizzling early-morning rain, my son Michael decides that we should search for the Silver Chief Mine. The night before, I told them about the prospectors and the chief's long freightliner canoe. We sit around the table finishing off our lunch. All of us are here, except for Patrick who had to work in town this weekend.

"Let's go and search for the Silver Chief Mine," says Michael.

CHIEF SEPASS ～ 89

"When?" asks Will.

"Right now."

"Now?" says Nora. "Dad, it's raining."

"Come on, guys," Kevin chimes in. "We can't let a little rain stop us."

"Yay, another expedition."

"Let's go."

"I'm coming too," says Molly.

"And me," says Joanna. "I missed the gravestone adventure. I'm not missing this one."

"I'm in," I say. "I talked to Kris Sanders last week. He gave me some instructions."

"What did he say, Ma," asks Kevin.

"He said to follow the boundary trail for about half a kilometre," I say, looking at my notes of the call. "When we find a square blaze cut into a large alder tree, we follow a track up the mountain."

"What's an alder tree?" asks Bronwen.

"What's a blaze?" asks Fiona.

"Don't worry, I'll show you what to look for," I say.

"We walked this trail when we went to look for the boundary markers two weeks ago. Looks like we missed that blaze," says Michael. "It must be hard to see."

"The trail switchbacks uphill three hundred metres," I continue. "We go across a small creek to a flat spot once used for a corral for packhorses."

"Packhorses? How'd they get way down there?" asks Fin.

"I think they were taken down the lake on a raft," I say.

"What's next, Mom?" asks Michael.

"Near the corral is a hole that has a rockslide across it. You have to climb over fallen rock but you can stand up inside. That's the mine."

"Hmm," says Joanna. "That's kind of vague."

"I've got it mapped out on my GPS," Kevin says. "We'll find it."

Michael and Molly, Kevin and Joanna, the grandkids, two dogs and I pile into the motorboat and head to the south end. I wish we could take the dugout. How perfect would that be? But only a few of us would fit and now the whole family is in a frenzy of curiosity. It would also take hours to paddle to the end of the lake.

When we arrive at the end, the wind has whipped up whitecaps and the waves are once again pounding the beach. Kevin and Michael follow the same routine, tying the boat to the deadhead offshore. Once we are on the trail to the border, we follow every side trail that might lead to the mine, but without luck. The clouds burst with a sudden shower, but underneath the canopy of cedars we manage to stay dry. We spread out along the trail and inspect every track that heads uphill. The kids quickly learn how to spot an alder and shout out whenever they see one. The leaves of the alder are egg-shaped with serrated edges and have distinctive veins running down the centre and sides of the leaf. The smooth ashy grey bark of the alder stands out against the brown stringy bark of the cedars. But no one sees anything that resembles the "square blaze" Kris described.

"Maybe this is something, " I call out. "I see a couple of red flag markers going uphill."

Michael follows the flashes of red. Kevin moves across from a higher position on the mountain. There are no switchbacks. It is straight uphill.

"Up here!" shouts Michael.

We rush up the hill, kids flinging themselves forward, each wanting to be the first inside the mine. We cross a creek and find what can hardly be called a clearing, but sure enough, boards are tacked to posts and hanging at odd angles. This must have been the horse corral.

"But where is the mine?"

Joanna finds a support beam across a rock face, but no opening. The kids scramble up and down the scree. I can't imagine hauling rock out of a hole in the mountain, packing the ore on horses and leading them down to the lakeshore on such a steep slope. It seems physically impossible.

Michael disappears from view, then a muffled voice calls out.

We follow the sound and stand on a mound of rocks, peering down at a dark gash three metres wide in the side of the mountain. Broken boards lie helter-skelter across the bottom. A glimmer of light flits across a dark pool of water. Michael's head appears, a smile on his face.

"This is it! The Silver Chief Mine."

I feel a jolt of excitement. Then a stab of concern. Is the mine safe? It hasn't been used in a hundred years. Should we go in? But the kids are already heading down.

"We'll be fine," I say to myself.

I abandon my concerns and slither after them down the rocks and into the mouth of the mine, heedless of the dangers. No boards shore up the sides, on all sides just a hammered rock face. The men would have used a chisel, called a "jack," placed in the rock face and hit with a sledgehammer to release a chunk of rock. Within a few steps, the mine is pitch black inside. Kevin and Michael tell us to wait. They round a corner and disappear out of sight. I turn on the light of my phone.

"Are we safe in here?" whispers Fiona. "I don't feel good."

"I can hear water dripping," says Fin.

"Let's look for silver in the rock," says Bronwen.

I shine my light on the walls and floor but no glimmers shine back.

Kevin and Michael return.

"The tunnel makes a few more turns but we didn't want to go any further. It could be dangerous. Let's head back outside."

Once we are out, we all breathe more easily.

"I wouldn't want to work in there," says Nora. "It's creepy."

"Yeah, but we found the mine," says Joanna. "Pretty exciting."

We head back down the hillside, slipping and sliding in the wet earth. The rain continues but we're oblivious, chatting about what the mine was like.

"It was so dark."

"Let's come back."

"Yeah, follow the tunnel."

"Not me. It could cave in."

"Let's go along the trail a bit further and see if we can find that square blaze," suggests Molly.

Only a few turns later and we find it: the faint outline of a square cut many years ago in the bark of an alder. The tree is large, maybe over a hundred years old. I wonder when the square blaze was cut?

"How'd we miss this?" says Michael, shaking his head.

I am thinking the same thing: what else have we missed, here and everywhere we look, the evidence of history passed by, overlooked. In my nearly seventy years here, I never saw what we've discovered in the course of the past month: a soldier's grave, a boundary line, and a silver mine.

"Let's come back with our Petzl headlamps and really explore that mine," says Kevin.

Chief Sepass and his partners spent three summers working the Silver Chief Mine, 1916 to 1919. To get to the mine, they travelled twenty-four kilometres over a wagon road along the Chilliwack River, rode another twenty-four kilometres over a horse trail to the north end of the lake, rafted eleven kilometres down the water, then switchbacked a kilometre up the side of a mountain. The mine was located about a kilometre-and-a-half further south than the camps the surveyors set up on Depot Creek sixty years earlier.

The mine was a horizontal hole cut into the mountainside. The horse corral was built on a patch of level ground. The miners fabricated a wooden channel to direct water from a nearby stream to a powerhouse, using pressure

Chief Sepass is standing on the top timber at the entrance to the Silver Chief Mine located at the south end of Chilliwack Lake. The mine was abandoned in 1919 after only a few years of minimal production of ore. Courtesy of Kris Sanders.

from water moving across the blades of a wheel to produce electricity. The prospectors constructed a log house along the banks of the Little Chilliwack River, with a lean-to kitchen and attached bunkhouse. After a few years of marginal silver yields, the mine was abandoned in 1919. I wonder whether we could find remains of the miners' camp buildings. When we go back to the mine, we should spend time searching along the Little Chilliwack.

Chief Sepass did not succeed in striking it rich. But his influence was felt throughout his community in other ways. He was known to be a great hunter. Although I haven't found any recordings of Chief Sepass's voice, I read a transcript of an interview conducted with Frank Malloway, a Stó:lō hereditary chief and descendant of one of the four original ancestors of the Ts'elxwéyeqw. In the quiet of the Stó:lō Centre, the old chief comes to life.

> When Chief Billy Sepass killed a bear, he went into a trance-like state. He talked to the spirit of the bear. He was thanking it as the bear was on its way to the Dead. A person had to have a lot of training to speak to an animal's spirit like this.

Chief Sepass was fluent in the languages of all the Indigenous peoples of the Chilliwack Valley, including the Ts'elxwéyeqw and Halq'eméylem languages. He also knew Chinook. He became the leader not only of the

Skowkale but all the other peoples of the Chilliwack Valley that eventually joined together as the Stó:lō Nation. He was a renowned orator, in the church and at the potlatch ceremonies that gave public recognition to the bestowal of hereditary names and the transfer of property.

The white missionaries were bewildered and appalled by the potlatch. In their eyes, the lavish gifting was wasteful. They lobbied the Department of Indian Affairs of Canada to ban the potlatch, and in 1884, Parliament enacted an amendment to the Indian Act making it illegal for Indigenous people to gather together in a ceremonial dance, funeral, marriage, naming ceremony, or any other kind of traditional event where gifts were given out. Despite this prohibition, potlatches continued in secret in British Columbia. Chief Sepass was the leading chief at the last great potlatch of the Ts'elxwéyeqw, held in 1916.

In 1922, the chiefs of the interior Indigenous peoples of British Columbia, including Chief Sepass, petitioned the government of Canada requesting that section 149 of the Indian Act prohibiting holding ceremonial gatherings such as the potlatch be amended.

The Indians of British Columbia prior to the advent of the Whiteman, had achieved a social and political organization very high for a primitive people. Owing to the fact that they were without writing, the only alternative was the administration of all their affairs with certain ceremonies, solemnities and certain publicity: this was done through what is commonly termed the 'Potlatch', or the proclamation of an event.

The Potlatch in its broadest sense is the gathering of those who are directly or indirectly concerned, either of one tribe, or more than one tribe, for the formal publication of some event whereby the status of an individual, or group of individuals was altered or transfixed. It included events such as birth, the giving of a name, marriage, divorce, election to Chieftaincy, and death. It included peace treaties, declarations of war, ratification of agreements, and transfer of property. It provided a record of the determination of any matter of importance.

It became the custom that the giver of the Potlatch would present gifts to those who attended, the more opulent the gifts the greater standing was held by the giver. The Potlatch covered every phase of the activities of the Indians, it was completely interwoven in their social life; it was the sole fabric and bond of the tribes and the families.

Charcoal portrait of Chief William Sepass Khal-agh-it-al, hereditary chief of the Skowkale First Nation, circa 1931. The renowned leader of the Skowkale First Nation, Chief Sepass lived over one hundred years, completing *Sepass Poems*, translated from the oral tradition, in his seventies. Courtesy of the Chilliwack Museum and Archives, 2002.095.001.

The petition was unsuccessful. The ban on the potlatch was not removed from the Indian Act until 1951, eight years after Chief Sepass died at one hundred and two.

Towards the end of my research into Chief Sepass at the Stó:lō Centre, I open the book, *Ancient Songs of Y-Ail-Mihth*, orated by Chief Sepass, recorded by Eloise Street and translated into English by her mother, Sophie White Street. I was saving this book for last. I wanted to know as much as possible about Chief Sepass before I read his poems.

In 1911, Chief Sepass met a disabled white girl sitting alone on a fallen tree on the shores of Cultus Lake, which drains into the Chilliwack River just south of Vedder Crossing. Eloise Street was seventeen and had polio. Chief Sepass sat down beside Eloise and recited a poem. He told Eloise of the loss of his people's culture, language and traditions. He had seen many ministers of the Methodist Church come and go, each reading from the Bible. The Bible was the constant, the written record. He wanted to have his poems recorded, translated and published, as a gift to his people, so they would know "what a very great people they had been." A book like the Bible. He requested her help.

> Chief Sepass saw in himself a symbol of the ending of a way of life. He had seen the White Man's books, and he wanted his songs in a book. He asked a promise, and the wording of it was a dedication: that the poems be put in a book so that Indians would remember their greatness for all time.

After four years of recitations, recordings and translations, Chief Sepass's poems were published in English. Because he had been trained as an orator from an early age, Chief Sepass was able to recite his poems from memory. The songs are part of an epic cycle beginning with the poem "The Beginning of the World."

The Beginning of the World

Long, long ago,
Before anything was,
Saving only the heavens,
From the seat of his golden throne
The Sun God looked out on the Moon Goddess
And found her beautiful.

Hour after hour,
With hopeless love,
He watched the spot where, at evening,
She would sometimes come out to wander
Through her silver garden
In the cool of the dusk.

Far he sent his gaze across the heavens
Until the time came, one day,
When she returned his look of love
And she, too, sat lonely,
Turning eyes of wistful longing
Toward her distant lover.

Then their thoughts of love and longing,
Seeking each other,
Met halfway,
Mingled,
Hung suspended in space . . .
Thus: the beginning of the world.

In the foreword to the *Sepass Poems*, the Honourable Steven L. Point, Lieutenant Governor of British Columbia, writes,

> Skowkale First Nation has a wonderful history full of very interesting people, one of whom is Old Chief William Sepass. His legacy has not only been his contributions as Chief but also his book, *Sepass Poems*. These old stories contain the very fabric that forms the basis of our cultural identity.

In the introduction, Chief Sepass's grandson, Gerald Sepass, adds,

Our grandfather's poems are from a long time before Europeans ever walked this land. He called them *Y-Ail-Mihth*, or *Uailmit, Songs of the Ancient Singer.* They were part of an epic cycle of songs that were traditionally recited at special gatherings and during the ancient 'Sun Ceremonies' that used to be held in Chilliwack every four years in pre-contact times.

The poems are ancient. They were passed on to our grandfather by his ancestors, and to those ancestors by their ancestors before them. Significantly, they represent the only work of traditional Halq'emeylem literature in translation.

Eloise Street recognized the duality of Chief Sepass's influence: "He was an Indian chief and a man of a very supreme knowledge of their culture, and on the other hand, he was a man who lived with White people who had completely different ways."

When he recited his poems, he was seventy. Not far from the age I am now. I feel like he and I are on the same path, wanting to preserve a tradition: him for a whole people, me in a much smaller way, for my family.

I started with the question of the connection between Poppy's dugout canoe and Chief Sepass. I still don't have a clear answer. Part of me hopes that Chief Sepass did not carve our dugout. Then maybe I wouldn't be faced with the dilemma. Should it be given back to the Sepass family? Or donated to the Coqualeetza Centre? Of course, this is not a decision I can make on my own. My brother Chris and the rest of my family will have a say.

I've always thought of the canoe as ours, a part of our history. But if it was carved by Chief Sepass, it rightfully belongs to the Skowkale First Nation. I wish I had certainty, but I just don't know who carved our canoe. Will I ever know for sure?

Charlie Lindeman

My mother must have known Chief Sepass. But it was Charlie Lindeman she talked about.

I walk into Crofton Manor, the extended-care facility in Vancouver where Mother has lived for the past ten years. It is the middle of July but the heat of summer is still in the wings waiting to step onto centre stage. It's not until August that we feel the incandescence of summer. I press the intercom to be buzzed in. The placard on the door admonishes, "Please make sure residents do not exit with you." The signs of aging envelop me: a man shuffling his feet, head bent over his walker to focus on the *fleur de lis* pattern on the carpet; a woman murmuring to a life-size doll, dressed in pink with a knitted cap, propped in her lap. The flat screen television in the lounge plays "The Hills Are Alive" from *The Sound of Music.* The residents are parked in rows, walkers and wheelchairs at their sides.

I find Mother slumped in her wheelchair in her room, curtains drawn. Light barely filters through the loosely woven fabric. Her gaze is distant, her hands loose in her lap. I approach her, touching her shoulder.

"Hi, Mum. It's me, Shelley."

I am preparing myself for the time when she will look at me without any glimmer of recognition. At the moment, her memory is elusive, but she still has moments of lucidity, a star in a cloudy sky appearing one moment and gone the next. This morning she startles, but when she looks up, her eyes brighten.

"When my eyes are closed, I can see the lake," she says, coming out of her reverie.

I keep her talking about the lake—the only topic that captures her attention. At ninety-five, she is a shadow of the vibrant woman she once was, but I still see vestiges of beauty in her face.

"Charlie and I carried fish to Lindeman Lake." A smile flickers across her lips.

She dozes off. Charlie Lindeman, originally from Wisconsin, came to live at Chilliwack Lake in 1911. For almost a dozen years, he was its sole inhabitant. Then Poppy arrived.

I remember the story of how they stocked Lindeman Lake. Mother was ten, Charlie fifty years older. Lindeman is an alpine basin two hundred metres up Post Creek from the Chilliwack River. Charlie and Mother netted fingerling trout in Post Creek, a kilometre south of Chilliwack Lake, then transferred them into metal buckets. They trekked up the three-kilometre trail, part of the old Indian Trail of a century before. They tipped the buckets into Lindeman Lake and watched the fish swim away. Mother always told this story with pride.

Chris pokes his head in the door. My brother is tall and slim, with hazel eyes set in a narrow face, a high forehead with grey receding hair cut short. He looks the most like Mother and Poppy, whereas I resemble Gran. He gives me a quick hug and kisses the top of Mother's head. She stirs awake and gives him a smile.

Chris has a blended family, five children altogether, ages twenty to thirty-three. When Patrick and I built our cabin in 1989, he took over our grandparents' log house, Cupola Lodge. Ten years later, he built an addition, adding more living space but keeping much of the original character. Keeping things the same is important to him.

We exchange news of our families, pausing to remind Mother who the children and grandchildren are. She doesn't follow the conversation but smiles at the banter between us.

Chris nudges me. I look over. Her head droops to one side. She has fallen back to sleep.

We tiptoe out, easing the door to her room closed.

"See you at the lake this weekend," I say as we get into our cars in the parking lot.

I think about Mother's life. To me, she is existing, not living. I know that she is safe and that this place has taken the burden of care off Chris and me. But this non-life is not for me. I would rather manage on my own than be cocooned in a care facility. I think Mother would feel the same, if she understood her circumstance. She was always fiercely independent.

I know this because, although I don't look like her, I *am* like her. I have always been independent too, wanting to live life my own way. As a girl, I was not trendy or fashionable. I quietly excelled at academics and athletics at York House School for girls in Vancouver. Achievement was more important to me than popularity. I was plain, with freckles, braces and dirty blonde hair. As a teenager, I wanted to see beyond my narrow world. At eighteen, I backpacked solo across Europe. At nineteen, I spent the summer living in Guadalajara learning Spanish. I joined the Canadian University Services Overseas Club while at the University of British Columbia, deciding in third year Arts that I wanted to teach in Africa. I was lucky to fall in love with a man who shared my vision.

I may be independent but I am not a patch on Charlie Lindeman, who was about as independent as a person could be. Charlie died five years before I was born so I never met him in person, but I feel I know him. He is a legend in our family: the trapper who introduced Poppy to the lake. His life before coming here is obscure. He was born Charles Otto Lindeman in 1869, in Pennsylvania, to parents of Dutch heritage. He died in Chilliwack in 1943 at seventy-two. He told Poppy that his family was poor: he left home in his teens because there were too many mouths to feed. He travelled south to Texas, then to the midwest, and ranged as far afield as Alaska. He never married.

Charlie Lindeman sitting on the porch of his cabin at Chilliwack Lake. Courtesy of the Chilliwack Museum and Archives, 2002.078.001.

During my summer, I come into the city for a day each week to catch up on errands and visit Mother. During my weekly visits, I prompt Mother to tell me more about Charlie. Gradually a picture emerges. Charlie smoked, either cigarettes that he rolled himself or else he had a pipe clamped in his mouth. He sported a black wide-brimmed hat, soft and well-worn. He sang to himself while he puttered around his cabin. He was slight with long, slender hands. He wore baggy pants cut from rough cloth, held up with suspenders. The pants were too long for his small frame, so he rolled them up at the cuffs. He wore baggy wool socks tucked into leather slippers that were cracked and shabby.

"When he smiled at me," Mother said, "I was his whole world."

What brought Charlie to Chilliwack Lake? I have pieced together the story from scraps of information. I found an article on Charlie published in the *Chilliwack Progress* newspaper in 2002. And the book, *In the Arms of the Mountains: A History of the Chilliwack River Valley*, includes some of Charlie's early history. Our neighbour at the lake, John MacLeod, told me stories of Charlie from when John was a boy.

In 1910, Charlie leaves Seattle, where he is living, and visits Chilliwack, renting a horse and buggy to drive around the area. He is taken with its beauty and determines to return. The following spring, at forty-two, he heads off from Vedder Crossing towards Chilliwack Lake, his horse loaded with provisions. Leading the way is Chief Sepass, then sixty-six. They follow the trail that winds through the Chilliwack Valley beside the rushing river, up ever-steeper inclines, switchbacking through the forest. The sun catches the sheen of the fir needles and reflects off the frothy white of the river's rapids. They set a good pace, reaching their destination at the end of the day.

Charlie sets up his tent in a clearing on the bluff overlooking the lake. From here, he can see straight down to the snow-capped mountains at the south end. Chief Sepass helps him set up camp, gathers some wood, then heads back down the trail. Over the next few months, Charlie fells the tall cedars that stand behind the clearing and builds a 4.5 metre by 9 metre cabin. He procures the basics—a wood stove, a few panes of glass for the windows—and makes a bed, table and chairs. His cabin is just one room. He builds a straight-backed bench on the porch where he sits to enjoy the view. He places a horseshoe above the door, nailing it upright to hold the luck. He builds a rough shelf on the porch railing. The next summer, he puts out pots of geraniums and snapdragons.

His cabin burns down. He rebuilds. He makes a raft but finds it unwieldy, so he hollows out a cedar tree, attaches a sail and lets the wind propel him down the lake. He bakes bread in an outdoor Dutch oven. He lives alone with his two Airedale dogs, Bob and Ben. He makes saddle packs for the dogs to carry their own food. He puts a paddlewheel that he made out of wood in a nearby creek to create power. He sets up a system of wires from the top of the bluff down to the water. A bucket is attached to one wire. When Charlie releases the bucket, it zings down the wire to the lake and fills with water. Then Charlie winches the full bucket back up the slope.

Ray Wells knew Charlie. I sit in my study and listen to Ray reminisce on tape:

> Yes, I knew Charlie very well, real well. He was a sociable sort of a fella. A man that was slow to make friends but he had lots of 'em. And he was a very reserved sort of a man. He was a kind of a man that you'd take to, you know. You'd like him when you first met him. He prospected and trapped for a living. He was very popular with almost all people.
>
> But in those days, of course, everybody had to go in on saddle horse or hike in. Not many of 'em hiked in because there weren't

too many bridges across the river. Or, of course, in later years, you could fly in. He grew most of his vegetables at Brown [Depot] Crick. The soil there was good. At the north end of the lake the ground was pretty rocky or pretty sandy.

He was a good man, really, none better.

Why would anyone settle upon such an isolated spot to put down permanent roots? Was Charlie running away from an unhappy past? Was he a recluse by nature? I choose to think of him as a man comfortable in his own skin and who appreciated the beauty of the wilderness he stumbled upon.

Solitude—a state of being alone—apart from other people. I am finding that I revel in being alone. I have time to think deeply. I am not distracted. I can focus. The balm of silence nurtures my spirit. I am discovering who I am and what's important to me. I savour solitude. It's what Charlie cherished. I do too.

As well as the cabin, Charlie builds a shack at the south end of the lake where he works a small garden in summer and sometimes overnights in winter when the weather is bad. Here he cuts cedar logs for the nine-metre dugout freight canoes that he uses for transporting the furs he traps. Charlie ekes out a living by trapping muskrat and marten. He sells the pelts for forty dollars apiece at the Fort Langley trading post, two days' walk from his cabin.

The trail to Chilliwack Lake goes by many names: Indian Trail, Hudson's Bay Brigade Trail, and Whatcom Trail. In the early 1800s, trappers travelled the Indian Trail between Fort Langley and Fort Hope to trade with the Hudson's Bay Company. In 1855, speculators from Whatcom County trekked the trail up from the United States to join the gold rush in the interior of British Columbia. In 1858, the trail led the United States and British delegations of the International Boundary Commission to their survey camps on Chilliwack Lake. Sections of the trail still exist, as popular hiking trails and part of the Trans Canada Trail system.

Where is the old Indian Trail? Another question for my notebook.

Poppy walked this trail when he first came to the lake, guided by Charlie Lindeman. Charlie is the link between Poppy and the lake. I know that Poppy worked for a federal water agency. He thought Chilliwack Lake might be a source of water for the Fraser Valley, and so he wanted to put a water gauge in the outlet of the Chilliwack River to record water levels.

But how did this lead Poppy to Charlie?

And in what year did Charlie introduce my grandfather to the lake?

I hire a researcher in Ottawa, Rachel Heide, to go over Poppy's employment files. She discovers that Poppy worked for the Water Power Branch,

BC Hydrographic Survey, which eventually became the Dominion Water Power and Reclamation Service. She turns up a wealth of files and sends me a disc to review. These are all employment files. I don't see anything that links Poppy to Chilliwack Lake. Rachel and I talk on the phone. What other files could she look up? Rachel suggests the British Columbia water gauge records. She finds a register with a record of water levels at the outlet of Chilliwack Lake. The first recording is May 1923. Finally, a date to hang my hat on. The water gauge was a priority for Poppy. If the date on the water gauge register was early 1923, the gauge must have been erected in late 1922. Poppy was a doer, like me: he wouldn't have waited to build the gauge. It's most likely then, that Poppy first saw Chilliwack Lake in the early summer of 1922. Until I find something that contradicts it, this is the date I am going to use.

Poppy meets Charlie in the spring of 1922 in Sardis, a small village at that time. Poppy asks around for someone to guide him to Chilliwack Lake. Charlie Lindeman is your man, everyone says. On one of Charlie's twice-a-year shopping trips to town, Poppy is finally introduced. Charlie is skeptical, not sure this posh gentleman can handle the trip. Poppy insists he's hardy and Charlie agrees to be his guide.

In late June, Poppy joins Charlie before daybreak at Bell's Farm along the Chilliwack River. Poppy brings a saddle-roll blanket and small duffel bag. In his bag, woollen socks and flannel shirt, a harmonica and a pocketknife.

I remember the knife. It had a deer-antler grip, with a nickel-silver bolster and brass liners, about six centimetres long. A lock-blade knife.

"The lock makes the knife safe," Poppy said to me when I was old enough to handle it.

Poppy never went anywhere without his harmonica. "You Are my Sunshine" and "I've Been Working on the Railroad" were his favourite songs. He'd play his harmonica in the evenings around our campfire, and we'd sing along with him. As teenagers, Chris and I joined him on the ukulele and guitar.

At Bell's, Poppy and Charlie saddle up the horses for the thirty-kilometre trek to the lake. As they climb the last hill before the benchland, they walk alongside the horses, swapping stories about spending time in the bush in the Yukon Territories. They have more in common than they expected. After a long day, they arrive at Charlie's cabin. Poppy looks down the lake.

"A piece of heaven," he says, a phrase he will say every time he arrives at the lake for the rest of his life.

They unload the horses and get a fire going. After a supper of beans heated in the tin washed down with a cup of coffee, they relax in front of the pot-bellied stove, stretching their legs. Slowly but surely the warmth of the stove dispels the chill that spins down off the glaciers surrounding the lake.

In the morning, Charlie and Poppy walk to the outlet of the Chilliwack River. The river is deep at this time of year, swollen with the spring rains and snowmelt.

"How can I measure the river's depth?"

"I could sink a pole at the edge of the river bank," says Charlie. "We could measure the river at that deep spot across the river. The current is swift but I could get over there in my rowboat."

"I know a way to get across," says Poppy. "I've seen a cable car on the Thompson River in the interior. I

The author's grandfather built a water gauge on the Chilliwack River that could only be accessed by a cable car strung over the river. The first recording from this gauge was in May 1923. Courtesy of Webb family.

could get one packed in here by horseback. And I'll send a roll of cable. Can you get someone to help you string the cable across the river and connect up the cable car?"

"If you get me the materials and a diagram, I can do the job."

"Okay. Once you get the water gauge in place, can you record the river's depth each day?"

"Sure can."

"I can pay you thirty dollars a month. You keep track of your hours and mail the log to me. I'll send you a cheque every month to your post box in Sardis."

"Sounds good," says Charlie. "I can use the money."

Maybe Poppy made it possible for Charlie to stay in the wilderness. This extra money might have made all the difference. And Poppy made it possible for me to have this "piece of heaven" too.

It is the third week in July and the boys' families are here for a week. The adults are up early and sitting on the dock ready to greet the sun, coffees in hand.

"Look how clear the 'Lady and the Lamb' is," says Molly.

Mount Redoubt is the highest peak at the south end of the lake. Just below its sentinel spires, a glacier forms two distinctive shapes. One looks like a woman in a bonnet and shawl holding out her hand. The other seems to be a small animal that she is feeding. "The Lady and the Lamb." Most people have a hard time deciphering the image, but once seen, the landmark inevitably becomes part of the conversation.

"The Lady and the Lamb is clear this morning."

"The Lady and the Lamb is in the clouds."

The grandkids straggle out, Fin and Roger from the boathouse, Fiona and Bronwen the last ones to leave the bunk room. Breakfast at the picnic table on the porch: cereal and yogurt with a big bowl of blueberries and raspberries.

The merganser family swims by.

"Oh no," says Roger. "I see only eleven babies. One is gone."

"I bet it's the little one that was always on the mother's back," says Nora. "It was the smallest."

"I hope we don't lose any more babies," says Fiona.

We feel protective towards the merganser family. Every day, we see them swimming along the shore in front of the cabin. It feels like they are part of the family.

"We'll keep an eye out for them as we walk to the river outlet," I say.

"Why are we going to the outlet?" asks Bronwen.

"I want to show you something."

"What?" asks Fin.

"You'll see when we get there."

"Let's pick up garbage on the way," says Will, ever the conservationist.

"Good idea," I say, stuffing a few plastic bags in my pocket.

We walk down the beach picking up bottle caps and candy wrappers, an occasional piece of glass dropped by campers and day visitors to the provincial park. The dogs race ahead, turn and bound back, checking that everyone is accounted for. Fin suggests a game.

"Let's go from here to the end of the beach jumping from rock to rock," he says. "If you touch the sand, you're out."

"Okay," says Roger. "I bet I can beat you."

Everyone joins in except Nora, who keeps me company. I guess, at fourteen, she's getting too old for games. I like the feel of her walking beside me, our heads at the same level. We are at ease with one another, chatting about school and summer plans. The oldest of the grandchildren, Nora blazes the trail. She was the first to swim to the deadhead, the first to drive the motorboat. If I want to know what's going on, I just ask Nora. I remember when I was fourteen. I thought I knew everything. Nora's quietly confident.

Will gives up on the game and runs up to me with another beer can to add to our stash of garbage. One by one, the others drop into the sand. Fin is the last one standing on the rocks.

We pass by groups of picnickers on the beach. When we come to the MacLeod cabin, we wave hello to John's daughter Susan and her family who are out on their porch.

At the boat launch, I lift the bear-proof lid on the park garbage can and stuff our collection inside. We leash the dogs with short ropes that I keep in my jacket pocket, then climb the trail from the beach up to the campground. The stretch of sunny and warm days has attracted lots of campers, many with RVs, tarps tied between trees to make a sheltered spot for sitting around picnic tables and fire pits.

Single file, we follow the trail along the bluff from the campground to the outlet of the Chilliwack River. We unleash the dogs. Roger and Fin throw sticks into the river and watch them swirl in the current.

"Don't do that," Fiona says. "The dogs might jump in."

"You're right, Fi. Not a good idea, boys." I stop and wait for them to gather round, which they do, like a collection of magnets.

"Here's what I want to show you. Look up at this tree. Can you see anything?"

"There's some rusty pieces of cable wrapped around the tree," says Roger.

"Yeah, and boards nailed in," says Bronwen. "They're pretty high up."

"Look across the river. Attached to that far tree, can you see more cable? That's where Charlie Lindeman built a water gauge to measure the water levels almost a hundred years ago. You remember me talking about Charlie Lindeman, right? The trapper who lived here before Gran and Poppy came to the lake?"

"How'd he get across?" asks Fin.

"A cable car and a pulley."

"Really? How?"

I tell them that an inch-thick cable stretched from the tree on the near bank to a tree on the far side. The cable car was suspended from the cable.

"I can remember going across with Poppy when I was a girl. It was just big enough for the two of us. I sat on the wooden bench at one end, with Poppy at the other end."

"How'd you get across the river?" asks Roger.

"Poppy stood up, grabbed the cable and pulled the car hand-over-hand across the river."

"How'd you get in the car, way up there on the tree?" asks Fiona.

"Charlie nailed rungs of a ladder and built a platform on each tree so we could climb up and get into the car."

"Did you know Charlie?" asks Nora.

"No. He died before I was born. Charlie recorded the water levels in the river for Poppy."

"What happened to the cable car?" asks Fin.

"I don't know. The car was still here when I was your age. One summer during university, it disappeared."

After Poppy's first visit to the lake with Charlie, he can't wait to return. Later that summer, he comes back on his own and sets up a tent beside Charlie's cabin.

The next day, he says, "I want to look for a place to build a cabin."

They walk along the shore together. Dense forest crowds close to the water. At the far end of the beach, they come upon a level clearing in a stand of old growth Douglas fir.

"This is the place," Poppy says.

He looks down the lake. Beyond a stretch of white sand, the dark water reflects craggy glaciered peaks.

The next summer, he brings Gran, Rowly and my mother, Frances. Rowly is seven and Frances is four. They set up tents in the clearing and Poppy plans his cabin.

Poppy picked the perfect spot. Did he have any inkling that this choice would be beloved by generations of his offspring? I wonder what he envisaged. A simple fishing camp or something more?

"Charlie, can you build me a cabin like yours? Just one room, a wood stove for heat, a door with windows on either side."

"Sure", says Charlie. "I can have it finished by next summer."

Charlie and Poppy tramp through the woods. They put tape around the western redcedars that could be used for the logs. Poppy returns to Vancouver, his mind alive with the vision of his cabin in the wilderness. During the early 1920s, guidebooks were published to encourage visits to Canada's national parks. Banff National Park was established in 1885, followed over the next thirty years by Yoho, Waterton Lakes, Jasper and Mount Revelstoke. The guidebooks portrayed these parks as sanctuaries of peace and tranquility, an escape from civilization—a concept that would have resonated with Poppy.

Using a single-head axe and cross-cut saw, Charlie fells each tree and trims the branches. His horse drags the logs to the clearing. Using only his horse and a system of pulleys, he raises logs almost a foot in diameter and up to thirty feet long. He places them on top of each other to a height of nine feet. The logs are notched at the ends with extended dovetail corners so that each log slots into the one below, a log construction known as "butt over top." The whole unpeeled log is used: the taper of the log requires that the narrow top end be matched to the wider butt end as the logs are placed on top of one another. When the structure is complete, Charlie stuffs moss chinking between the logs and nails saplings over top to keep the natural cushioning in place. Around the cabin, he builds a seven-foot wide porch. The roof slopes out over the porch. Charlie hand-cuts cedar shakes for the roof. Three steps lead up to the porch at the front door. The only windows are on either side of the door.

The following year, Poppy arrives at Charlie's by horseback a week ahead of his family. Charlie greets Poppy with a smile.

"Your cabin's finished. Want to see it?"

Charlie and Poppy work long days to get the cabin ready for the family. They build a table and two benches for meals on the porch, wooden bed frames and a small table to hold a washbasin. The wood stove brought in by packhorse is transported from Charlie's to the clearing by boat and set up in the middle of the cabin, connected to pipe that rises in a curve up through the roof. The following week, the horse wrangler, Bridge Bailey, arrives with Gran, Rowly and Frances and a packhorse train loaded with supplies for the new cabin.

Every summer, Poppy lets Charlie know when they will arrive and Charlie is waiting at the top of the trail to greet them. Charlie keeps watch over Cupola Lodge through the winter to make sure that everything is ship-shape for their return.

One day when the men are sitting at the table after dinner at Cupola Lodge, Charlie says to Poppy, "I'm thinking that I should own my land."

"I've had the same idea myself," says Poppy. "Why don't you do a diagram setting out the dimensions. I'll talk to a lawyer I know in Vancouver."

Charlie Lindeman in dugout canoe on Chilliwack Lake, circa 1922. Charlie was a trapper and prospector who introduced the author's grandfather to Chilliwack Lake. Courtesy of the Chilliwack Museum and Archives, PO466.

In 1985, after we moved back to Vancouver and I took over the management of our holding company, Cupola Estates, Dad handed over three cardboard boxes of files. Most of them related to routine matters such as property taxes, water licence fees, annual filings with the Companies Registry. There were no diaries and few photos, but I found some gems. A sketch by C. O. Lindeman dated September 1, 1925 showed his cabin sitting on a rectangle of land, approximately seventy-six metres by a hundred and fifty metres. Nine years later, title was transferred to Charlie from the British Columbia Department of Lands. Charlie pays seventy-five dollars for his 14.8 acres. Poppy buys his first parcel of land at the lake for the same five dollars an acre.

A few years later, Poppy visits Charlie at his cabin with a proposal. Poppy sits on the high-backed bench while Charlie perches on a three-legged stool made out of a burl from an old fir tree. Poppy says he has a plan for breeding marten and muskrat. Charlie tells Poppy about the time he raised marten. He built pens for them and, once they were fully grown, he carried them to town in cages on his packboard. As he walked, the marten would run end to end in the cage. In the steep sections of the trail, Charlie would lose his balance. He and the marten would fall over on the trail. Poppy laughs along with Charlie imagining this sight. He knows that Charlie finds tending his trapline two thousand feet above the lake hard going at fifty-six.

"Let's go into business together to breed muskrats," says Poppy.

Among Poppy's papers, I find a 1927 hand-written agreement describing the breeding of fur-bearing animals to produce pelts and skins for sale.

Lindeman would manage the business, would purchase 10 pairs of muskrats and put them in pens on a breeding ground located approximately three miles from Chilliwack Lake. Webb would provide funding for purchase of muskrats, wirenetting for pens, an Evinrude motor and obtain a lease of the grounds.

I don't know if they went ahead with the fur-breeding business. I find no record of a lease.

Gran and Poppy spoke of Charlie with great respect, but an element of mystery hung in the air. Charlie chose a solitary life, living his last thirty-five years in the wilderness of Chilliwack Lake. During the winter months, Charlie was captive in a world of white. Snow blanketing the cabin, ridges of ice on the lake, blustering winds and baying wolves. Surprisingly, he was not quiet and taciturn. He was a raconteur with a gift for story-telling. If a visitor dropped by, he was regaled with tales long into the night.

When Rowly finishes high school, Poppy decides that he should spend the winter with Charlie at the lake. He is nineteen. Gran argues against the idea, but Poppy is adamant, and so, early in the winter of 1934, Poppy delivers Rowly to Charlie's cabin with a duffle bag of clothes and books. Rowly is to stay with Charlie for five months. Charlie has built a makeshift sleeping cabin for his visitor. Rowly, tall, slender, his angular face an aesthetic version of Poppy's, stoops to enter the cabin.

The winter is harsh. In January, temperatures dip to minus thirty degrees Celsius. Snow falls steadily, reaching a record two metres. Rowly accompanies Charlie on his trapline, at first struggling, then mastering the art of the snowshoe.

Rowly's snowshoes hang on one wall of our cabin. The wooden frames are a metre long, with a tapered tail and rawhide webbing, a leather binding to attach a shoe. We still use them if we come to the cabin in the winter.

As is his custom, Poppy leaves Charlie his gramophone—Rowly's one connection to civilization. Listening to Mozart, Rowly can escape the confines of the cabin and be transported to a world of lights and glamour where he feels at home.

The gramophone was portable, in a polished mahogany case, with "Victrola" etched in a brass plate on the top. It was Poppy's pride and joy. Gran kept a bottle of lemon oil on the dresser. Once a week, I smoothed a few drops of oil over the gramophone's surface and polished it until it shone.

"This is precious," Poppy warned me. "Treat it with kid gloves."

Putting the record on the gramophone was a precise performance. Lift the lid until it latched in place. Place the record on the turntable, but first wipe it clean with the blue velvet dust cloth. Crank the handle clockwise ten times to wind the mainspring, no more, no less. Release the brake lever to start the turntable spinning. After the turntable has revolved several times and picked up speed, lower the sound box gently on the smooth outside rim of the revolving record. Then place the needle into the sound wave groove. Poppy told me it revolved at seventy-eight revolutions per minute. The music began and another world opened.

When I am a girl of eight, Poppy says to me "I have a surprise for you."

He puts a new record on the gramophone, from the musical *Oklahoma!* I sit transfixed. I play it all day, learning the words, singing along to "Oh, What a Beautiful Morning" and "The Surrey with the Fringe on Top." I can still remember the words. Whenever *Oklahoma!* is playing at Theatre Under the Stars in Vancouver, Patrick and I take the grandkids. We all sing along.

Rowly follows those same precise movements to play his gramophone records during the winter of 1934. The only recordings he has are symphonies by Mozart and Beethoven, and an opera in German.

Charlie sets traps for snowshoe hares, one of their food staples during the winter. Rowly checks the traps daily. He watches the hares change from a patchy brown and white coat in November to fully white by January, camouflage to match each stage of groundcover through the winter. The hare's only distinguishing mark in winter is the black tufts of fur on the edge of its ears. Rowly comes to recognize the hare's hind print pressed an inch into the snow, which, along with the tail print, looks like a miniature snowshoe. He approaches each trap with apprehension. He hopes the hare is dead. If not, he has to steel himself to club it over the head.

In April, Charlie walks Rowly out to Vedder Crossing where he is met by Poppy. Five months have gone by.

When he returns to Vancouver, Rowly is a silent presence in the house. He and Gran go for walks together, huddle over cups of tea. He takes a course in design at the Vancouver School of Art. He and Gran discuss their mutual love of art. He goes to live in Victoria and frequents Maynard's art auctions. He waits for his chance to escape.

The war is his ticket out. In 1939, Rowly joins the Royal Canadian Air Force. He is twenty-three. He becomes a rear gunner and is shot down three times over the English Channel. He is diagnosed with battle exhaustion and sent to recuperate in the English countryside at Little Coxwold Grove, Faringdon, Oxfordshire. Geoff Berners, whose family owns the manor, takes Rowly in, treating him as the son he never had.

In 1947, Gran takes a train to Halifax and boards the ship S.S. *Aquitania* for England. Rowly meets her at Southampton. They haven't seen each other for eight years. Rowly drives Gran in Geoff Berners' black Rolls Royce to Faringdon.

"My British family," he says, introducing them.

Gran and Geoff Berners take walks in the garden. Rowly is happy in England and wants to stay. Geoff offers to set Rowly up in business in London. Gran agrees.

"Come back every few years, if you can," she says to Rowly. "Frances wants to see you, and so does your father."

Rowly remains in England for twenty-seven years. He opens, in succession, a men's clothing store, a toy soldier store, and an antique store, all in Picadilly in central London. He returns as promised every few years.

As a young adult, I could not understand Poppy's icy indifference towards Rowly. I understand better now. When I look at photos of Rowly as a boy and as a young man, his sexual orientation leaps off the page. Poppy turned away from Rowly and focused his attention on my mother, Frances, who became the son. Acknowledging a son who was gay would have been unheard of in Poppy's circle of friends; Rowly was an aberration in his eyes.

After Poppy dies, Rowly moves back to Vancouver permanently. He is fifty-five years old. Finally, he is able to live openly as a gay man. For the next twenty years, Rowly phones Gran every morning at ten and they talk about everything and about nothing. They go to the podiatrist together. They go to lunch, both of them impeccably dressed. Rowly sets up an antique store in South Granville, importing paintings and furniture from England. He becomes a collector of British Columbia Indigenous art, with the brooding forests of Emily Carr on his walls and Haida black-banded baskets of tightly woven spruce root on his bureaus. Visiting Rowly's apartment is like going to a gallery. Every inch of wall space and table surface is covered in art. I wonder why he focused on Indigenous art. He obviously recognized the power in the artistic expression. Did he value its essential link to the natural environment? I wish I had asked him when I had the chance.

Rowly only spoke of his winter at the lake once to me. He was in his late seventies; I was a young mother with two boys. I asked him what it was like spending the winter in the wilderness. He said that it was hell, but that he was saved by the kindness of Charlie.

"Charlie was a saint for putting up with me."

He didn't say more. I guess Charlie and Rowly were kindred spirits, both needing to escape the world of constraint.

What was it about the lake that answered a need in Charlie? It wasn't just the isolation. Charlie had lived in places just as remote. The magnetic pull of the lake goes beyond its beautiful setting. Here a person can become still, at one with its beauty. Charlie had found his spiritual home. But the solitude wasn't for everyone. It wasn't for Rowly. But it was for me, and it was for Poppy.

POPPY

The screen door bangs and there's a rustle of newspaper, the clunk of the lid of the wood stove, the hiss of a match struck, a crackle as the fire catches. A murmur of voices comes from the lean-to kitchen attached to the side of Poppy and Gran's one-room log cabin. The air is crisp, with the sharp scent of fir needles, and now, a faint whiff of wood smoke. I am nine, lying in my bed on the back porch, Chris on the cot behind me. Slowly I open my eyes. The shadows of green branches flicker across the slatted rafters of the porch. A shiver of cold in the air makes me snuggle back into the blankets.

"Shelley, Chris, time to get up," Poppy's voice calls out.

"Wake up," I say, tugging at Chris's bed covers. When Poppy calls, we hop to it.

I pull on khaki shorts, a faded T-shirt, sweater, and runners and smooth the sheets and blanket. I put my Nancy Drew book, *The Secret of the Old Clock*, on the wooden butter box that holds my clothes. The smell of sizzling bacon and toast draws me into the kitchen. Another day at the lake begins.

We sit down for breakfast in our allotted places: Mother and Dad, Chris and I on benches along the sides, Gran and Poppy at either end of the table.

"Shelley, do you want to come fishing with me after breakfast?" asks Poppy.

I nod and smile, happy to go with him. I clear my tin plate and cup and gather the soft canvas bag with the fishing gear and the short-handled net that hangs on the log wall. Poppy goes ahead. The canoe was pulled up high onto the beach last night, as usual. Poppy pushes it part way out into the water and steps in to settle on the seat at the far end of the canoe. I run across the sand to join him. I like joining Poppy in his daily trolling ritual.

"Let's get a couple of trout for dinner," Poppy says.

I kick off my runners and step into the water. A shiver runs up my legs. I lean forward to put the bag and net in the canoe, then push the boat off the beach. I step in, careful not to wobble. I like the way the breeze makes ruffles on the water, but not so much that it will be hard to paddle.

Our dugout canoe only has two seats, one at each end. We face each other, Poppy at one end and me at the other. The way the seats are arranged, only one of us can paddle and he gestures to me that I should pick up the paddle

The author's grandfather in a wooden rowboat on Chilliwack Lake. Courtesy of Webb family.

that is lying in the bottom of the boat. His black felt cap slouches at an angle, his body at ease on the cracked wooden seat. He opens the fishing bag and brings out the tackle box. I watch to see which lure he selects to attach to the line. Everything my grandfather does is of deep interest to me: he is like a long drink of water I can never get enough of. I dip my paddle into the water, trying not to make any noise. I hear a clunk and flinch. My paddle has hit the side of the boat. Did Poppy hear it too?

"Shelley, be careful," Poppy admonishes. "Any noise will scare away the fish."

"Sorry, Poppy."

He frowns and turns his attention back to choosing his lure. He sets out the trolling line. I concentrate on keeping my paddling stroke steady, using the J-stroke he taught me, careful not to bang the side of the boat.

I run my hand along the outside surface of the canoe and feel the feathered pattern. Poppy told me that the canoe was hollowed out of a cedar tree many years ago. He showed me the tool that would have been used, called an adze. Poppy's face has a pattern of lines, as if he, too, was carved by an adze.

I look at his mouth to judge his mood. If it is relaxed, with an upward curl to his lips, he is happy. If his lips are tight and pressed together, he is not. This morning, Poppy's face is stern. I try hard not to bump the side of the canoe again. His line jerks and he reels in a rainbow trout.

"That's a good size," he says. "A couple more and we'll have enough for dinner. Paddle us along the shore in front of Cupola Creek. We'll see what luck we have there."

He catches five in all, four rainbow trout and a bull trout.

"Let's head for home."

His eyes crinkle and his face relaxes. He smiles. I like it when he smiles at me.

A lot of water has flowed under my canoe since then. I'm sixty-six now, my grandfather has been dead for forty-one years, but in this summer of discovery, I am determined to learn more about him and his connection to Chilliwack Lake. If it weren't for him, our family wouldn't be here. I have found out when he came to the lake but I don't know why he chose to stay.

What made him choose this remote location for a fishing cabin?

As I note this question in my notebook, I realize that the more questions I answer, the more I seem to add to the list.

I know very little about Poppy's early life. I need to do some digging. I sign up with Ancestry.ca and start searching on the website. Poppy's birth certificate lists his father's and his mother's name. From here, I find his siblings. I troll through a number of search categories: births and deaths and marriages, immigration and travel, census and voters' lists and military enlistment. I come upon a photo of his father, Christopher Morrison Webb, as a middle-aged man. He wears a full beard and moustache and a three-piece suit. Aside from the stern countenance, I don't see any other similarities to Poppy. I turn up an image of a headstone for my grandfather's father, his mother, and two of his sisters.

Poppy was born Christopher Everest Webb in 1887 in Granton, a village twenty-six kilometres north of London, Ontario. John MacLeod once told me that Poppy made a point of emphasizing his middle name. Poppy seemed to have an affinity with Mount Everest, officially named in 1865 for Sir George Everest, a surveyor general of India. I imagine Poppy would have celebrated the news in 1953 of the first official ascent by Sir Edmund Hillary and Tenzing Norgay. It is appropriate, I think, that the man with this moniker as his middle name eventually had a mountain named after him.

Poppy was the youngest child of eight. There are eighteen years between him and his oldest sibling. A few names that I uncover in my searches sound familiar: Olive, Daisy, Clarence and Fannie. I never met any of them. Although he carried the name of a mountain, his sisters called him "Christie." I can't imagine this. Such a babyish name doesn't fit with my image of Poppy. But it's appealing, showing tenderness for the youngest of the family. My search shows up a United States immigration form for his brother Rowland, a surgeon, who settled in Grand Rapids, Michigan, at twenty-three. His sister Jessie, a nurse, married a doctor, also from Grand Rapids, Michigan. Another relation, who was called "Spider" Webb, was also a doctor. Clearly a family of high achievers. Poppy must have felt a certain pressure to live up to the accomplishments of his older siblings.

Poppy attended the University of Toronto, which seems unusual for the youngest child of such a large family; he graduated with his Bachelor of Science just after the turn of the twentieth century. His mother valued education and encouraged Poppy to go as far as possible in school. In fact, the commitment to higher education is a thread that is strong in the Webb family. Poppy encouraged my mother, his daughter Frances, to take her Masters of Social Work, unusual for a woman during the early years of the Second World War. I realize now why no one blinked an eye when I said I wanted to be a lawyer. Education would have been a sticking point with Poppy's son, Rowly, who did not go to university. I'm not even sure that Rowly finished high school. I can remember Poppy saying to me, "Your studies are the most important thing. Do well in school and you will succeed in life."

Poppy's father ran a general store and his mother taught school, both of them pillars of the Granton community. When Poppy was in grade school, the family moved to Toronto, where his father expanded his business, opening several more stores. During one of my parents' many house moves, I ended up with a box of old photographs. I brought the box up to the cabin with me at the beginning of this summer. I have looked through the images many times over the past weeks. They have no captions, no names scrawled across the back, so it's impossible to know who's who. I find one with a passel of children: the smallest, in the centre, in short pants and knee socks, could be Poppy. He is staring ahead, serious, hands folded in front.

The author's grandfather conducting an engineering survey for Canadian Northern Railway in British Columbia, circa 1912. Courtesy of Webb family.

After graduation in 1907, Poppy left his family and travelled out west. Men were being lured to the west with promises of land. But he wasn't a farmer. He was an engineer. He joined survey teams in the backcountry of British Columbia and in the Yukon. He spent months in the bush, his pack roll strapped to the back of his horse, a way of life nothing like his father's genteel stores or his brothers' surgeries. Riding, hunting, setting up camp, he learned to fend for himself. In the photos of him, even those taken in the bush, he is always impeccable, never dishevelled in the least. So what was the appeal of this rough-and-ready wilderness life? What was he hoping to find?

Rachel Heide, my researcher in Ottawa, looks into Poppy's employment with the federal government and sends me two discs. The record starts with a letter offering Poppy a job in 1913 and ends with the notice of his retirement from civil service in 1951. There are over a thousand documents: letters, cablegrams, requests for leave, salary negotiations, travel itineraries, and meeting agendas. I set to work.

The first correspondence is a job offer from the federal Water Power Branch in Ottawa. The letter is addressed to Christopher Webb at the Walhachin Hotel. Walhachin is a railway station town in the Thompson River valley, west of Kamloops. At the time, Poppy was working as a resident engineer with the Canadian Northern Railway, stationed in Walhachin.

Walhachin had a romantic history. Established in 1908, its advertisements as a land of bounteous fruit orchards attracted aristocratic settlers from England. The town had amenities unheard-of in other towns: entertainments by travelling musicians, current affairs coverage in a local newspaper, tennis courts, a swimming pool and elaborate gardens. The Walhachin Hotel boasted two wings with a second floor promenade. The elegant dining room enforced a strict dress code. My dapper grandfather would have been at home. Unfortunately, the exaggerated promise of the ads was a sham. Walhachin, in fact, was set in one of the driest regions of British Columbia. A hurriedly-constructed irrigation system was inadequate and the orchards failed. Many of the first townspeople returned to England during the First World War to fight for their country. By the mid-1920s, Walhachin was a ghost town. Skeletons of elegant houses stood like tombstones on the arid land.

Poppy was there during its heyday but he didn't stay to watch its demise. His job offer was confirmed by letter in the fall of 1913:

> I beg to say that your application of the 20th of August last to Mr. R. C. Swan, Chief Engineer of the B.C. Hydrographic Survey, for a position on his staff, has been transmitted to the Department and approved, and that authority has been given for the utilization of your services on this work, at salary of $135 per month, with travelling and living expenses when in the field. .
>
> I am unable to say at the present time just when you will be requested to report to Mr. Swan, but expect to be able to do so within a short time. When you do report to him you will be expected to give him your most loyal and efficient service; all promotions and increases in salary depend on his recommendation.

Poppy replied to Mr. Swan with a handwritten note. I recognize his penmanship. I stop and stare. His writing is neat and legible, letters flowing and connected, slanting to the right in a firm expression of competence. His sloping letters bring back a flood of memories of Poppy's notes at the lake.

Poppy wrote instructions for everything. Every can in the gashouse had a tag attached, written in his precise script: "Gas for 5 horse motor, mixed with oil"; "Kerosene for lamps," "Kerosene for fridge." Each key in the ring that hung on the hook beside the cabin door was connected by a piece of string to a round disc identifying its lock location. Details for lighting the fridge, connecting the water and cleaning the coal oil lamps were taped to the inside door of the high cabinet along the back wall of the cabin. In everything, he was exact, leaving nothing to chance or mistake. Poppy's notes are still thumbtacked into the inside door of the tall wooden cabinet in the old cabin; "How to fit copper tubing" and "Directions for hummingbird feeder." A detailed sketch of the water pipes labelled "Webb's Water Pipes - scale 4ft to 1 inch."

The next correspondence is from Kamloops, in central British Columbia, where Poppy moved to take up the position of assistant engineer with the Water Branch.

I switch back and forth between his employment records and the documents I found on Ancestry.ca. Here is his marriage certificate, dated August 19, 1914, just after the beginning of the First World War. He married Frances Ferguson, the daughter of the hotelier in Savona, a small town sixteen kilometres east of Kamloops along the Thompson River. They were wed at Christ Church Cathedral in downtown Vancouver. Gran and Poppy started their married life in the Kitsilano neighbourhood of Vancouver, along the southern shore of English Bay.

Poppy's employment files show him to be a dedicated worker, often not taking the full leave he was entitled to, especially in his first years. He was in charge of identifying new water sources for Vancouver and the Fraser Valley. This is what led him to Chilliwack Lake. His search for water was not a success, but he found something else. A summer retreat. He returned and brought his family with him. But why this place? He must have seen dozens, hundreds of lakes in the course of his work. I wonder if he was inspired by the challenge: a place so remote that it took a full day's ride on horseback to reach, a place no one else had discovered, a place untouched by civilization. Or maybe he was drawn by the isolation, the chance to get away and be alone. Maybe the man who hunted and camped in the Yukon was the real "Christie," not the Vancouver engineer.

None of the documents, plentiful though they are, give me a glimpse into his heart. And I never thought to ask him while he was alive what it was about the lake he loved. I can only guess at his motivation. The Poppy I knew was most at ease quietly paddling the dugout. He sought solitude. I think the desire for tranquility was the magnetic pull.

I imagine the first time Poppy brought his family to the lake.

It is July 1923. Poppy finishes packing the black Dodge touring car and calls to his family to get in. Gran is apprehensive of such a long, arduous excursion with a four- and seven-year-old in tow. My mother and Rowly sit wide-eyed in the back seat. Since the previous summer, Poppy has talked of nothing but Chilliwack Lake. The family will camp for two weeks in utter wilderness, no other humans for hundreds of miles except for Charlie Lindeman. The children have never met a trapper. They have never been in the mountains. Poppy has packed minimal supplies: basic foodstuffs and utensils, an axe and hatchet, fishing gear, bedding, two tents, waterproof matches, candles and a small cook stove. He will borrow Charlie's boat and fish for supper every day.

The family commandeers two of the thirty-eight rooms of the recently completed Royal Hotel for their overnight stay in Chilliwack. With electric lighting, steam heating and telephone conveniences, the hotel rivals the best to be found on the coast. They drive the next morning to Bell's farm, at the halfway point between Vedder Crossing and Chilliwack Lake. Charlie Lindeman and Ed Bell have the six-horse pack train ready. Frances sits in the saddle in front of Poppy, Rowly on the back with Charlie, Gran on her own horse. Gran is wearing navy twill trousers and a wide-brimmed felt hat. Her belted afternoon dress, matching gloves and purse were left behind: not camping attire. A couple of horse wranglers lead the loaded packhorses. The morning dew dissipates as the string of horses follows the trail. The group takes breaks, giving Frances and Rowly a chance to stretch their legs. They stop for lunch by the river's edge. Gran has packed picnic lunches for everyone. The sun beats down on their heads as they ride through the lodgepole pine in the valley, then the trail climbs into the cool tall forests of Douglas fir and western redcedar. Gran, Frances and Rowly droop with the setting sun as they arrive at Charlie's cabin. They dismount and Poppy leads Gran to the edge of the bluff overlooking the lake.

"Here it is," says Poppy, looking at Gran with anticipation. "What do you think?"

"It is beautiful, Chris. Truly."

The family beds down on the floor in Charlie's cabin. After breakfast, Charlie piles them into his boat, taking several trips to ferry their supplies to Poppy's chosen spot down the beach. Here they pitch their tents, haul water, set up a makeshift kitchen under a tarp strung between two trees, dig a latrine and settle into camp life.

Poppy's patience must have been tested, camping with two young children in such an isolated place, his family relying on him for everything, even the next meal. But he perseveres.

"Next year, we will have a cabin to live in," he says as they say farewell to the lake at the end of their two weeks.

"You bet," says Charlie.

When Gran, Rowly, and Frances arrive the following July, Charlie and Poppy have the single-room cabin ready. The family sleeps on the wrap-around porch on bedsprings set in cedar frames, Gran and Poppy in front, Frances and Rowly on the side. The cabin has an attic accessed by a steep set of stairs and a trap door. Here, the bedding is stored for the winter in a large wooden box lined with tin. A small cast-iron stove heats the cabin. On the front porch sits a table and two benches for their meals.

In the years that follow, Charlie builds several outbuildings: a wood-shed at the back, a fish shed attached to the woodshed and a gas shed some distance down the beach. The crowning glory is the outhouse, set off in the woods, down a trail, and away from the cabin. It has two holes, including a small one for children, and a window looking out into the trees. In nearby Cupola Creek, Poppy finds a natural pool with fast-flowing water. He wedges a length of canvas hose between two rocks, dipping the mouth deep into the pool. Gravity sends water down the hose to the cabin. Initially, the water flows freely onto the ground. It will be a few more years before the hose is hooked up to a tap at the side of the cabin. One summer, an Elmira cook stove arrives, strapped to the back of a horse. A lean-to kitchen is added to one side of the cabin: a rough shelf, hooks for pots and mugs and wall brackets for plates and lids. The stove is on rollers so it can be pushed back and locked into a storage closet for the winter.

Eventually a second room was added to the cabin. When was the kitchen built, I wonder? I can't find any information in the cardboard boxes or any of the other papers. I riffle through the photographs and find one of me standing in front of the cabin beside a woodpile. I look about six. The cabin has one room. I find another photo, of me brushing my teeth at a washbasin on the porch. I am maybe seven: the cabin is still just one room. I keep looking until I find a photo of the cabin with two rooms: I am holding up a fish by the gills, the cabin in the background. I am ten: I remember those shorts. From this simple sleuthing, I deduce that the second room was added some-time between 1955 and 1958, which meant that Poppy and his family sum-mered in that single-room cabin for over thirty years, while the family grew. Gran and Poppy, Mother, Dad, Chris and I all shared that same small space

The author's grandparents and family on the porch of the cabin on Chilliwack Lake, circa 1933. Courtesy of Webb family.

for all that time. Maybe Poppy wanted to keep life simple. Maybe he didn't like change.

I always knew Poppy as mercurial; one minute charming, the next minute ill-tempered. Now I think I understand this dichotomy. Being the youngest in a family that valued success and social standing, he was dedicated to excellence, a perfectionist whose temper flared when perfection was not achieved. His anger rarely fell on me, but it caused pain to Gran, his son Rowly and daughter Frances.

I pull up the canoe and come around the cottonwood trees between Poppy's old Cupola Lodge and Sandpiper, the new Pan-Abode cabin my parents are building. I hear raised voices. Poppy is pointing at holes cut at the bottom of the Pan-Abode walls. Beside him, Mother's shoulders are hunched, her arms folded tight across her chest. I stop, not wanting to approach. I am thirteen and curious, so I eavesdrop.

"Why are these holes here?" Poppy shouts.

My mother's voice is agitated but firm. "They are for ventilation."

"They'll let in mice."

"The builder says we need them. He nailed screens over them."

"You have to fill them in. Right now."

I can hear Mother crying.

"Why couldn't you build a log cabin like ours?" Poppy says and stomps away.

I am upset with him and I feel sorry for Mother. Why does he insist on having his own way? Why does he get so angry?

When we are still living at Cupola Lodge and it is cold and rainy, Chris and I sleep on metal cots in the attic. We each use a chamber pot at night so we won't have to disturb our grandparents by climbing down the ladder into their bedroom and out to the outhouse. We waken to the noise of breakfast being prepared in the kitchen below. I dress quickly and straighten the bed covers. The pale morning light ekes its way through the tiny window at one end of the attic. Chris pulls on his pants. His leg gets stuck and he jerks it forward, knocking over his chamber pot. Yellow liquid spills through the cracks in the rough floor, dribbling down into the kitchen.

"Chris, get down here right now," yells Poppy. "You've made a mess!"

Chris looks at me, scared.

"It's okay," I say. "I'll come down with you."

My heart sinks. I know he's in serious trouble. This isn't the first spilled chamber pot this summer. Although Chris is only eight, Poppy expects him to be careful. And Poppy gets madder each time. I wish Poppy could excuse Chris. He's just clumsy. He doesn't do it on purpose.

As I continue reading through Poppy's employment file, something strikes me. There are at least two requests for medical leave every year, some for as long as two weeks. The conditions vary: influenza, bronchitis, pneumonia. All lung ailments.

Was Poppy sickly as a child? I don't know. I remember that Gran and my mother both fussed over him, always worrying about whether he was warm enough. They didn't want him to catch a cold. He pulled a knitted nightcap snug over his head when he slept on the porch, a practice he continued even after he and Gran moved inside the cabin. He donned a wool cap during the day, too. The top button of his shirt was always done up. He often wore a wool vest, even in summer. He never talked about illness, but I remember a constant undercurrent of worry about his health.

I didn't think about any of this until I read his many medical leave requests. I review another government form. The box was checked for military service, with a footnote that Poppy was discharged from military service in 1916. Was that due to respiratory weakness? He never did serve overseas. I recall now that Mother had asthma as a child and the family moved away from the fog of the harbour to the more inland neighbourhood of Kerrisdale for the sake of her breathing. Don't illnesses run through families? Of our six grandchildren, five of them have had respiratory issues. They've outgrown them, but maybe this weakness in the lungs stems from Poppy. I haven't inherited this particular flaw, but what other traits do I carry in my DNA? My desire for solitude, my passion for wilderness, my obsession with orderliness, my urge for perfection: perhaps these, too, came from Poppy.

Cupola Lodge was Poppy's domain. Every detail was created to his specification. Poppy made the decisions and solved the problems. All the systems were of his design. The gravity canvas hose system for obtaining water from the creek. The container for tins of butter lodged in the creek to keep them cold. Holes dug as root cellars for vegetables at the edge of the clearing, protected from rabbits by a wooden box set firmly on top.

Poppy always had to be in control. His world had to be perfect and in order. But Poppy couldn't control everything. Not his health. Not the lake. He brought his family to live in the wilderness. He tried to bring order: for every tool in the shed, he painted an outline on the wall; for every machine, he wrote instructions; and for every part that might break, he had a spare. But lakes flood, streams burst their banks and fire destroys. Did he rail against nature or yield to it? If he is anything like me, he probably did a bit of both. I like to tame the space around me but I also like the thrill of being on the edge. Standing on the beach with a storm raging around me. Knowing I am powerless. Maybe in this we are alike.

If Poppy wanted to be master of his own destiny, why did he wait so long to buy the land? I do a British Columbia Land Titles Search and obtain a copy of the title. I am surprised. Poppy built the cabin in 1923 but the title was not acquired until 1929. He was a squatter for six years. This seems out of character for a man I knew as a stickler for order. I see that the title was originally in the name of Ronald Stockton, barrister and solicitor of British Columbia. I remember hearing Mr. Stockton's name: he was a friend of Poppy's as well as his lawyer. Less than a year later, Mr. Stockton transferred the land to Frances Ellen Webb, my Gran.

I finally figure it out: Poppy did not buy the land because, at the time, a federal employee was not allowed to buy land from the federal government. I obtain copies of the titles, seven parcels in all, purchased over the course of thirty-five years. Mr. Stockton would buy a parcel, then transfer it into Gran's name. I wonder why Poppy didn't buy the land in Gran's name from the beginning as women were allowed to acquire land in British Columbia, although women were not "persons" until 1929. The last piece was bought in 1957, which brought the family holding to sixty-eight acres. In 1958, the land was transferred to a family holding company, Cupola Estates.

I am grateful that Poppy bought the land when he did. We live on less than five acres, but the extra land gives us a buffer against encroachment. We are privileged to have this piece of wilderness. We now summer on the shores of a pool of water that sits like a jewel in the middle of a huge provincial park. I remember Poppy saying to me once, "This is not yours. You hold the land in trust for future generations." I say the same thing now to my sons, to my grandchildren. But we also have received this land in trust from past generations. We owe it to them to take care of this place.

Poppy revelled in living in a log cabin in the woods, isolated, relying on his own skills. He was also at ease with city life. He was charming and sophisticated, a member of the venerable Vancouver Club, where he joined Vancouver's business elite for lunch. His manners were impeccable, his pocket handkerchief always matched his tie, his shoes were polished every evening.

In the summers, Poppy was a wilderness recluse; in the winters, he was steadily climbing the ladder of success at the Water Power Branch. In 1925, he became British Columbia's District Chief Engineer, responsible for establishing water resource policy in British Columbia and the Yukon Territory.

I find a press release issued October 18, 1950:

> The Helmand River Delta Commission has been appointed by Afghanistan and Iran to render an advisory report to the two governments on problems of mutual interest connected with the Helmand River. The Commission consists of Senor Francisco Dominguez of Chile, Mr. Christopher E. Webb of Canada and Mr. Robert L. Lowry of the United States. The three commissioners, who are highly qualified and well-known experts on river water problems, were selected from a panel of internationally prominent engineers.

I remember Poppy talking about Afghanistan. One of Gran's prized possessions was a powder compact, silver with inlaid enamel painted in an intricate flower and star design, a gift Poppy had brought back from Afghanistan. She gave it to me when I moved to Vancouver in 1985. It now sits in a place of honour in our living room in North Vancouver.

When he returned from Afghanistan, Poppy suffered two severe attacks of pneumonia. In the spring of 1951, he requested early retirement, which was immediately granted. In his retirement notice in the *Vancouver Sun*, he was lauded as "Mister Water."

I had no idea of the esteem in which he was held in Canada and around the world. I was three when Poppy retired and he rarely talked about his work after that. I wish I'd asked him about his time in Afghanistan. I would love to have heard about the complexities of conducting field studies and diplomatic negotiations in a country so different from Canada. When I had the opportunity, I wasn't interested. Now that I'm interested, he's no longer here.

My clearest memories of Poppy are from when I was a girl at the lake. Once the float plane dropped us at our beach, the six of us were on our own for four weeks, with only an occasional visit from the scattered few living nearby. This could have made our family close-knit, but that was not the case. We were not a family who communicated easily. Our summer world was orderly, with Poppy as the quartermaster. Chris and I knew our duties and tried to accomplish them to Poppy's satisfaction. It was an idyllic life with an undercurrent: don't rock the boat. Poppy had standards, and he was quick to let you know if you were not measuring up. I walked on eggs, trying to gauge his mood.

Exploring in the woods one day, Chris and I find two Douglas firs set side by side in the middle of a clearing. We want to build a fort. We settle for a lean-to. We take a saw, a hammer and a tin can of nails from the shed. Dad helps us nail a top beam between the two trees. We cut saplings and lay them down the sides, covering them with fir boughs. It's not watertight but it's ours. Chris, who is nine, names it Fort Red Pine, after a rusty-needled pine tree at the edge of the clearing. Every day after morning chores, we head out to Fort Red Pine. One day, Poppy comes to see what we are up to. He is not happy. We cut a trail to the fort. We took tools out of the shed without his permission. We are wasting nails. I stand with my head down. Chris scuffs his feet in the dirt.

"Sorry, Poppy," we mumble over and over. When he finishes raging, he stomps off.

A few days later, he comes back. Chris and I glance nervously at each other.

"I have something for your fort," he says.

He's carrying a folded tarp.

"If you put it on the ground, it won't be so damp. But you have to put it back in the tool house at the end of the summer."

"Thanks, Poppy," we say in unison.

He smiles as he leaves us.

Poppy is resourceful. The summer that I am six, wasps are everywhere, caught in the jam on my toast, struggling in the water in my glass. Poppy strings up the head of a fish he caught that morning on the branch of the birch tree at the side of the cabin. Under the fish head he places a pail of water. The wasps gorge themselves on the fish, then sated, drop into the pail of water and drown. The pail fills with wasps and we eat in peace. The following year, Poppy installs netting across a section of the porch to create a protected dining area.

The only time Poppy takes a back seat is when the kitchen is built. This is Gran's domain. Gran chooses pale yellow for the counters and a central island that has deep drawers for flour and sugar, Gran's baking counter. She sews curtains for the windows, white with yellow and blue flowers. A varnished cedar table sits in the front window with benches on each side for meals and games. A coal oil lamp hangs over the table from a metal chain that can be lowered for lighting. The lamp's glass shade is decorated with a painted meadow scene, pale blues and greens. The wood stove is connected to a hot water tank. She designs a kitchen sink with water taps, hot and cold.

When Poppy runs the plumbing to the kitchen, he adds a shower stall on the back porch. No bathroom: we're still using the outhouse. The original room of the cabin is now Gran and Poppy's bedroom. My parents sleep on the side porch with Chris and I on the back.

Sometime during its first winter, the new kitchen is christened by an intruder. Charlie Lindeman arrives to discover the hinges ripped off the kitchen door, the drawers of Gran's pastry table torn out and flour and sugar strewn all over the floor. Charlie cleans up the mess. When we arrive in the summer, we gape at the claw marks on the doorframe and the battered metal drawers of the pastry table.

We never entertain another bear, but Poppy wages war with other critters that come uninvited into our cabin. For mice, Poppy uses the tin-can-over-the-pail method. Each morning, furry bodies float belly-up in the pail.

For packrats, Poppy devises something different. The size of a small squirrel, with big ears and a long bushy tail, packrats make nests out of chewed bits and pieces of paper and cloth. The animal's pee has a penetrating smell, a bit like the stench of a skunk. We are on the lookout in the spring and fall when packrats migrate into the cabin in search of a cozy home. One whiff of a packrat and Poppy sets a live trap, a rectangular metal cage with metal mesh bottom, sides and top, and a metal handle on top. My job is to drown the rat snarling up at me from the cage. Poppy tells me not to be squeamish: rats are smelly and make a mess of the cabin. They deserve to die. So I march the cage to the dock and drop it into the water, just deep enough to drown the packrat. I turn my head away so I don't have to look at the struggling rodent. When the cage is still, I bring it back and Poppy tosses the body into the wood stove, the fire stoked.

Then there are the bats. When I am ten, my dad has a woodsman from the valley spend a couple of weeks with us to build a boathouse for storing boats in the winter and sleeping in the summer. Chris, Mother, Dad and I are the builder's helpers, picking up bits of wood, sweeping up the shavings. Almost as soon as the boathouse is finished, bats take up permanent residence in the eaves of the boathouse porch. Poppy tells us to ignore them. "Bats are good," he says. "They eat mosquitoes, flies and moths."

I am not convinced. I have visions of the flying rodents getting snagged in my hair. But Poppy's word is law. We leave them alone.

Nora, Fiona, and Roger are spending the last week of July with me at the cabin. Over breakfast on the porch, we watch a dozen hummingbirds crowding around our feeders, screeching and diving at each other, the males flashing their red throats at the females. They arrive in late April and gather here until late August when they head out on their 3,200-kilometre flight to Mexico. Now at their peak, I change the feeder water at least twice a day, keeping the whirlwind of winged bodies satiated.

The otter shepherds her three kits on their morning foray to the river outlet to fish. The kits stay close to the mum. They slip up on our dock, rubbing themselves on the wood planks. When they slide back into the water to continue on their way, the smallest of the kits is left behind. Its high-pitched cry pierces the air. The mother swivels her head and with a sweep of her tail, turns around and comes back. The cries continue until the mother returns to the dock. Reunited, the kit joins the lineup and the family continues on its way.

After breakfast dishes, we walk over to Cupola Lodge, Poppy and Gran's cabin taken over for many years now by my brother, Chris and his family. As we climb the steps to the porch of Cupola Lodge, Roger, always observant, stares intently.

"Rani, is that a fish drawn on the log?"

The image is so faint I can hardly see it. We go closer and see several outlines of fish drawn on the porch logs.

"I remember watching Poppy draw this one on the wall. I was nine," I say, pointing to the largest image on the log. I read the inscription: "July 28, 1957. Rainbow Trout. 21 inches."

"Here's another one," says Nora: " 'August 4, 1954. Rainbow Trout. 16 inches.' I wonder who caught that one?"

"Probably Poppy. He caught all the big ones. He liked to keep a record."

The fish images on the logs are rainbow trout, ranging from forty to fifty-three centimetres. I had forgotten that the drawings were here. We don't have stuffed fish or animal heads on our walls. Patrick likes to preserve images in photographs. He has taken spectacular pictures of the otters, bears, owls, deer, Canada geese and hummingbirds. These images hang on our walls and grace our photo albums.

I point out the area of the porch that used to be screened in, then walk around to the back of the cabin, showing them where Chris and I used to sleep on the back porch.

"Weren't you afraid of bears?" Fiona asks.

"No, we liked sleeping here. But a bear did come up on the porch once."

I wasn't terrified, I was intrigued.

I lie on my cot on the back porch, Chris just a few feet away. I am eight and he is six. It is early morning, the sun not yet up, but the sky lightening above us. I awake to the tang of recently-cut cedar kindling stacked in the woodshed. Squirrels chatter as they chase each other up the giant fir that stands just off the back of the porch. They make such a racket. They seem to be saying to me "Get up, you lazybones!"

I detect an unfamiliar noise. The hair on the back of my neck stands on end. I hear a snuffling and a ponderous step. I look up to see a bear at the woodshed lumbering towards the porch stairs.

"Chris," I hiss. "Wake up."

His tousled head lifts off his pillow. "What?"

"Come with me. No questions, just come."

He sees how serious I am. He slides out of bed and we slip around the far end of the porch, into the cabin. I shake Poppy's covers, "Poppy, a bear's on the porch."

Poppy sits up in bed, "Keep still. Stay here, you two."

The bear is now on the side of the porch that leads into the kitchen. We crouch down and peer through the windows, wondering if the bear is going to come crashing in the kitchen door. I look over at Poppy. He puts his finger to his lips. My heart is pounding. I peek out. The bear is ambling along the porch and down the front steps. Poppy planted an apple tree beside the porch and this year it has produced a few apples. The bear rears up on its hind legs, pulling down the branches and swiping at the apples. It pulls itself into the tree, a huge Winnie-the-Pooh in a spindly shrub. We hear the sound of cracking limbs, and the bear tumbles to the ground. He shakes himself, then meanders down the beach and into the forest, out of sight, like an actor in a play I'm watching, not threatening but definitely entertaining. We laugh.

When Poppy told the story, and now when I tell it, we call it "The bear that came searching for apples at Chilliwack Lake."

When I am thirteen, life changes abruptly. In 1961, the Chilliwack Lake Road is completed, linking Vedder Crossing with Chilliwack Lake. Now we can drive into our cabin. We can come and go as we please. The next year, my parents build their own cabin, a Pan-Abode that Mother names "Sandpiper" after the birds that often bob on the lakeshore. Poppy, at seventy-four, doesn't adapt well.

"I don't know why you have to build a new cabin," he grumbles. "We were just fine the way we were. Your house is so big. And it's not made of real logs."

I suspect he was upset that we would no longer be together in the same cabin.

Gran and Poppy start coming to Cupola Lodge less and less. Poppy is no longer comfortable making the long drive. When I get my driver's licence, I sometimes drive my grandparents to their cabin. Poppy relinquishes the reins in other ways, too. Dad takes over the paperwork for the property, filing company returns, paying property taxes, responding to mail. Poppy lets Mother and Dad make more and more of the decisions.

I spend time with Gran and Poppy in the city. During my grade twelve year, my parents decide to build a new house in Vancouver. We become a family of nomads. For six months, I live with Gran and Poppy. Life is well ordered in their house. Breakfast on the table at seven sharp, porridge cooked overnight and served with cream and a dollop of maple syrup. Poppy drives me to school. He still goes downtown every day, to work as a consultant for the hydro industry. He arrives home by 5:30, pours himself and Gran a gin and tonic, and opens a bottle of Coke for me. Dinner is served at six, in the dining room, with the best silver and china. There's always a pudding for dessert. He likes to discuss current affairs at the table and asks my opinion. He quizzes me on how my classes are going, what teams I'm on. Poppy and I play a game of cribbage while Gran does the dishes, both of us engrossed, calling a "muggins" if one of us misses a point. He leaves me to watch my favourite television show, *Peyton Place*, at nine: no remonstrations about the frivolous show I am watching. He is happy to have me in his house. I can speak up now without fear of his displeasure. I have grown into a confident young woman. And he has mellowed.

I don't see as much of Gran and Poppy through my university years, although I live at home and attend the University of British Columbia. I am away most summers, with jobs and travelling. I work as a bank teller and clean cabins at Jasper Park Lodge. I squeeze in a few weeks at our cabin where I can.

Patrick and I meet on the lawn in front of the library during my first week of first year at the University of British Columbia. In my fourth year, we spend our mid-February reading week at the lake. My parents are away and we have snuck in on our own. Snow blocks the access road, so we leave Patrick's car on the Chilliwack Lake Road for the week. Poppy phones me a few days after we return. The Royal Canadian Mounted Police have just called him to report that a car was found abandoned on Chilliwack Lake Road. The licence plate belonged to a Patrick O'Callaghan. They were concerned that there might have been a break-in. Poppy assured them that it was a friend's car and thanked them for their diligence. I listen with a sinking heart, waiting for the reprimand.

"Next time, let me know if you go to the lake on your own," he says, "especially in the winter."

"Okay," I say surprised and relieved. "Thanks, Poppy."

He never tells my parents.

Poppy is hospitalized in Vancouver with respiratory failure the first week in September 1974. I give birth to our second son, Michael, on September 10, 1974, in Toronto. Patrick takes photos of the baby and mails them to Vancouver. They arrive at the end of September and my mother takes them to the hospital. Poppy holds them in his hands and smiles.

"His name is Michael," she tells him.

Poppy dies a week later on October 2, 1974. One life ending just as another begins.

In my mind, I am at the dining room table in the old log cabin. Poppy sits in his place at the head of the table. The light of the coal oil lamp casts a glow across his face. He lifts his eyes and smiles at me. This is how I like to remember him: the light of his love shining on me.

Now that I am the same age he was when I was a girl, my judgment of him has softened. I didn't understand his rejection of Rowly. I didn't like seeing him get angry with my mother, his daughter. Now I see the grey rather than the black and white. I see, too, that I have many of Poppy's traits: the attention to detail, the joy in teaching the next generation. Are they flaws? Maybe. Or maybe they are what keep us together.

GRAN

I LUV U MOM, carved into a piece of driftwood by ten-year old Kevin and given to me for Mother's Day thirty-two years ago, hangs on a wall in the cabin, one of a long line of hand-made gifts.

It is late July. All of us are together this weekend, celebrating Patrick's birthday. My granddaughters and I go for a walk in the morning to find decorations for the birthday cards they plan to make. The girls wear shorts and halter tops and have bare feet, feeling the summer's heat even first thing in the morning. The forest floor is a mosaic of green and yellow. We find a multitude of mosses: club, yellow-green and feathered; stag's horn, plump with multiple tendrils, intertwined like emerald pipe cleaners. Nora picks the feathery club-moss. Fiona keeps searching, looking for something more unusual. Bronwen chooses multiple shades of green.

As we walk around the Sandpiper, Chris's oldest children, Alex, Blake and Hannah, are sitting on the porch finishing their morning coffee.

"Hey," says Blake. "I've been wanting to ask you about your expedition to the soldier's grave. How was it?"

"Awesome!" "Cool!" "Amazing!" Chorus the three girls.

"We found the headstone," says Bronwen. "Right in the middle of a swamp."

"Yeah, there was Devil's Club everywhere," says Fiona.

"And we walked down the river at the end of the lake," says Nora. "With bears around the corner!"

"It sounds pretty exciting," says Blake. "I definitely want to go exploring."

"Yeah, let's do it," says Alex.

"Well, Kevin has the grave's location on his phone's GPS, so he can give you directions," I say.

"What are you up to now?" asks Hannah.

"We're looking for decorations—leaves, moss, twigs—for birthday cards."

"Oh, it's Uncle Pat's birthday!"

"Why don't you guys come over for drinks tonight? About 6." I say.

"Sure, we'd love to come," says Hannah. "See you later."

When we come in the door of Creekside, the girls place their forest treasures on the games table at one end of the living room. They get out construction paper, coloured pencils, glue, scissors and a box with snippets of ribbon saved over the years. They pull up their chairs and start working on their cards for their grandfather—Poppy to them—another family tradition carried forward.

"I'm going to write a poem," says Nora. "Poppy likes jokes, so maybe a poem with a joke."

"I'm going to glue the twigs in a cool design," says Fiona, who adds a certain flair to everything she does.

"I'm not sure what I'm going to do," says Bronwen, staring down at the materials. She likes things she does to be perfect.

"Who started the idea of making presents anyway?" Nora asks.

"Chris and I always made presents for my Gran's birthday. Maybe the tradition was started before that, I don't know."

"What kinds of things did you make?" asks Fiona.

Gran's birthday, on August twenty-first, became our annual end-of-summer celebration. The gift planning started almost as soon as we arrived at the cabin in mid-July. Each year, Chris and I would try to think of something unique. One year, I etched the words "Cupola Lodge" on a piece of driftwood with a penknife. I burned the etched letters using the sun and a magnifying glass so the black letters stood out against the varnished wood. The sign still hangs on the wall of Cupola Lodge. Another year, I searched until I found the perfect bracket fungus, one with a ridged brown top and milky white underside. I carefully removed it from the fallen tree and drew a picture of the cabin across its smooth, unblemished underside.

Gran's favourite gift was a concert: Chris and I put our own words to "Five Foot Two, Eyes of Blue" and sang it to her, both of us strumming our ukuleles.

> *Five foot two, eyes of blue, oh what those two eyes can do*
> *Has anybody seen our Gran?*
> *In every game, she wins her fame*
> *Croquet, Yogi—drives us insane*
> *Has anybody seen our Gran?*
> *Well, she can bake a cake, wield a rake*
> *Anything you please*
> *Cut down trees, exterminate bees*
> *Everything for her's a breeze*
> *Today's the day, for all to say*
> *Gran you sure do lead the way*
> *Has anybody seen our Gran?*

I know the date of Gran's birthday, but not the year. I can't recall her maiden name except that it starts with an F. I look back at Poppy's marriage certificate. Gran was a Ferguson. I log into Ancestry.ca and search for Frances Ferguson. Up comes Gran's birth certificate. Born Frances Ellen McMillan Ferguson on August 21, 1893 in Savona, British Columbia. I remember that she was the youngest of four children—one older brother and two older sisters—so like Poppy, she was a youngest child. Her father, Adam, was born in Ireland and immigrated to Canada in 1862, five years before it became a nation. Her mother, Sarah, was born in Oregon, and married Adam when she was twenty-four. He was thirty-nine. They lived in the interior of British Columbia in the small town of Savona, a mid-way stop on the Cariboo Wagon Road between Barkerville in the north and Yale on the Fraser River. Savona was bustling during the Gold Rush of the 1850s. Adam was the proprietor of the Savona Hotel, where gold-hungry prospectors stopped to rest and where the Pony Express paused for food, lodging, and a fresh horse before continuing on its way.

I find the death certificate for Gran's mother, Sarah, who was thirty-nine when Gran was born. She died when Gran was only four. Gran had no memories of her mother and her father, who was fifty-four when she was born, was mostly absent from her life. I remember Gran saying that she was raised by her older sister, Rita.

Gran talked about going to a boarding school in Yale, the name sounded something like Hallowe'en. I look up schools in Yale in the early 1900s and find All Hallows School, the first boarding school in the interior of British Columbia. An online article called "The Rise and Fall of All Hallows School in Yale, BC, 1884–1920" provides the basic history.

All Hallows School was started in 1884 by the Sisters of the Anglican Church for the education of Indigenous girls. Six years later, daughters of white settlers and clergymen were included. The school became known for its high standard of education. Its violin orchestra performed across Canada. It had a tennis court, hockey and croquet grounds. A nearby creek was flooded during the winter for a skating rink. Gran was six when she was sent to All Hallows School. I read that schoolgirls waved branches of maple leaves when the Duke and Duchess of York visited the school in 1901. I imagine Gran as one of those girls. Eight years old, old enough to remember, although she never talked about this event or about the school, certainly not about how lonely she was, how she felt to be almost an orphan, how she felt about sharing her school with Indigenous girls.

I think about her, a motherless little girl. I imagine her getting on the stagecoach at Savona, her trunk tied to the roof, driver sitting up front, six horses in harness. Were any of her sisters with her, or was she alone? The Cariboo Wagon Road was completed in the late 1800s from Yale to Barkerville to service the gold fields. The teams of horses were changed every thirty kilometres at stations along the way. Gran's trip from Savona was about a hundred and sixty kilometres and would have required an overnight stay. I hope someone was with her. She would have travelled this route four times a year, coming home only at Christmas and the summer. When I think of myself at the same age, I lived a safe and predictable life in Vancouver, walking two blocks to school, coming home for lunch. I have new respect for the spunk Gran had as a girl.

At thirteen, Gran went to St. Ann's Academy in Victoria on Vancouver Island to complete her schooling. I find documents describing both All Hallows and St. Ann's on the BC Archives website. On Monday afternoon, I drive into Vancouver with a plan in mind. I visit Mother, stay overnight, then drive early the next morning to the Tsawwassen ferry terminal just south of Vancouver. I am going to spend the day at the BC Archives in Victoria on Vancouver Island. I drive off the ferry and along the highway into downtown Victoria. I had printed a map back at home and find the BC Museum easily. I park in the museum lot and bring out my case with my computer, notebook and research notes.

From the parking lot, I walk through Thunderbird Park beside the museum. I stop to look up at each of the eleven totem poles. I study a Haida pole with depictions, from top to bottom, of an eagle, mountain hawk, whale, human head and beaver. Another that catches my eye is a Kwakwaka'wakw pole, with a thunderbird, grizzly bear, beaver, and wild woman of the woods holding her child. Intricately carved and painted, the poles stand nine to twelve metres tall. Once skyscrapers themselves, the totems are now dwarfed by the downtown office buildings. I walk by a replica of a longhouse from the eighteenth century. I wish there were a Ts'elxwéyeqw pit house, so I could see how the Indigenous people at the lake lived.

Beside the museum is the BC Archives. I descend the concrete stairs to a walkway surrounded by water. The cedar doors of the archives are carved with animals, each a symbol in Haida culture. Intrigued by the intricate carving, I pause to run my hands over the chiselled wood before I open the door.

A security guard greets me at the front desk, signs me in and runs down the list of rules. No pen, purse, briefcase, or jacket is allowed in the research room. I place my belongings in a locker. I feel stripped down. This is so much more formal than my visits to the Chilliwack Archives and Coqualeetza Centre where I've spent so much time this summer and where I feel at home. I nod to a second guard inside the research room. I fill in request slips, one per item. I complete the first few quickly and get them in the queue as there is no retrieval between 11:30 and 1 pm and the room is crowded. My name is called out. I am given instructions. Documents must be laid flat, only one out on the table at a time. Originals must be handled with gloves.

I open a folder containing notes concerning British Columbia place names collated by A. G. Harvey between 1906 and 1950. Here are some answers to my confusion on the multiple names for places at the lake. Lindeman Lake, formerly Silver Lake, was officially named in 1951 by the BC Geographical Names Department in honour of Charles Lindeman, trapper and prospector in the area for thirty years on application by Walter Gilbert of Lindeman Lodge, Chilliwack Lake. Depot Creek was also officially named in 1951 to recognize the international boundary survey camps along the creek that was formerly called Brown Creek (to commemorate the death of United States scout Michael Brown). I find that Paleface Creek was named in 1939 even though we were calling it Kokanee Creek well into the 1950s. I can find no reason for this change of name.

I pick up the book, *Indian Education in Canada, Volume 1: The Legacy* and flip to chapter six: "Separate and Unequal: Indian and White Girls at All Hallows School." At the time, this was the only boarding school in Canada that enrolled both Indigenous and white girls. In the beginning, the girls were taught together and learned the same material, but soon, the two were segregated. The Indigenous girls no longer learned reading, writing and arithmetic. Instead they were trained in laundry, cooking and cleaning. They became the servants. They did the work that kept the school running, while the white girls prepared to be accomplished wives and mothers. Two visions emerge of All Hallows School. Most accounts report that Indigenous girls were treated like second-class citizens, relegated to demeaning work and excluded from academic opportunity. On the other hand, some considered the school enlightened. How did Gran feel about this treatment of the Indigenous girls in her school? She never said a word.

Recently, I finished reading the chapter on the history of residential schools in the *Final Report of the Truth and Reconciliation Commission of Canada*. This segregation was repeated endlessly across Canada's residential schools in British Columbia, on the Prairies, in Ontario and Quebec and the Atlantic provinces. This heritage affects us all. I am affected in the abstract, but now I know that my grandmother was a witness to the cruel and

Sisters and pupils lined up on a pathway outside All Hallows School in Yale. The author's grandmother attended this boarding school for both Indigenous and white girls from 1900 to 1906. Image C-08200 courtesy of the Royal BC Museum and Archives.

discriminatory treatment of girls with a different skin colour. I would like to be able to talk to Gran now and hear her perspective.

I go out for a break at noon. In the entrance, I catch a glimpse of a sign over a door across the hall. The Sisters of St. Ann's Archives. I walk over for a closer look. The office is closed, only open by appointment. I make a note to look for an email address. Back in the archives, I open an 1858 pamphlet, the original prospectus for St. Ann's Academy. The school's mission was to "impart to young ladies the benefit of a good moral and domestic education, accompanied with the knowledge of the various branches of elementary training, together with those which constitute the higher departments of a finished education." Run by Roman Catholic nuns, St. Ann's was the first boarding school on Vancouver Island. The original schoolhouse, built in 1844, was relocated to Thunderbird Park. When I leave at the end of the day, I walk over to the one-room log house, now painted white, about the size of Gran and Poppy's old cabin. I wonder what the school looked like when my grandmother enrolled fifty years later.

That night, I look up the St. Ann's website and send an email asking if they have a record of Frances Ferguson. A few days later, I receive a reply:

Ferguson, Frances, age 13, Savona BC entered St. Ann's Academy in September 1906. She took music as an extracurricular. Her sister Katie (age 16) joined her 23 Oct 1907. In 1907, Frances started taking drawing lessons and by 1908 they were both taking water-colour classes.

When Poppy died just after Michael was born, we were living in Toronto. The family didn't have a funeral for Poppy as he had requested that we hold a family memorial service at the lake instead. This memorial was planned to take place the following spring when all the family could attend. But I couldn't wait: in late November I flew from Toronto to Vancouver to visit Gran.

Mother picks newborn Michael and me up at the airport. I break down and cry as soon as I see her. I ache, feeling the loss of Poppy like a severed limb. She gestures to me and I place Michael in the crook of her arms, tucking his blanket around him. He is awake and stares up at her.

"Look at him," she says. "He looks like Poppy."

"I want to go and see Gran right away," I say.

"I thought you would," says Mother. "She wants to see you too."

Mother pulls into their driveway and I hop out, retrieving Michael from his car seat in the back. I walk up the stairs to the front door. I still think of this as Gran and Poppy's house. I take a deep breath and open the door.

"Gran? We're here."

She comes down the hall to greet me. The two of us hug each other, tears flowing down our cheeks. Michael is squished between us. He starts to squirm, then lets out a bellow.

"Let me get a good look at him," Gran says, smiling through her tears.

I bring him into the den and place him in her lap when she sits down.

"I like the name Michael," she says.

A pale light filters through the French doors leading to the backyard. The limbs of the cherry and plum trees create a dark silhouette against the grey November sky. I can't believe that Poppy will never hold Michael. I start to cry again. My tears fail to fill the bottomless hole in my heart.

Mother has come in behind me and fetches the tray Gran has ready in the kitchen.

"Are you doing okay without Poppy, Gran?" I say. "I miss him."

"I do, too. But now I have you here. And this beautiful baby boy."

We have a cup of Earl Grey tea and Gran's oatmeal raisin cookies, using her best silver tea set and china. She wants to know all about the baby, how he's feeding, whether he's sleeping through the night yet. Just before I go, Gran asks me to come with her to the attic. Mother takes Michael into the kitchen while she cleans up the tea things. Gran and I bring down three paintings.

"I wanted you to see these," Gran says. "I painted them at school when I was seventeen. Maybe you'd like to have one?"

"Gran! I didn't know you were a painter. I'd love one. These are beautiful. Why didn't you keep painting?"

She gives an imperceptible shake of her head but no answer.

The paintings are water colours, although the paint has leached out of the paper leaving the images faded. All three are copies of French Impressionist paintings in fashion in the early 1900s. *Frances Ferguson* is written in a clear cursive hand in the bottom right-hand corner. They now hang in the living room at the lake. My favourite is a rendering of Pissaro's *Landscape at Osny*. It hangs beside the chair in the corner where I like to sit and read.

Later, I ask Mother why Gran stopped painting. She tells me that when Rowly was eight or nine, he became interested in drawing. Gran bought him sketchpads and graphite pencils. He and Gran went to art galleries. He loved beautiful objects and would pore over art books at the library. Poppy became more and more upset. He decided to curb Rowly's interest in art. He signed Rowly up for tennis lessons at the Vancouver Lawn Tennis Club. He removed all art supplies from the house. He forbade both Gran and Rowly to paint.

I am shocked to hear this. It pains me to think of her giving up her art all those years ago. It pains me to know that Poppy insisted on this sacrifice. How did Poppy feel enforcing this command? I know he loved Gran. Was he distressed, to cause her pain? What did he hope to gain? Did he think he could change Rowly? I wonder why Gran submitted. But what else could she do? In those days, wives did not defy their husbands. I can't imagine being in such a one-sided relationship when my life with Patrick is so filled with give and take, resolutely based on mutual respect.

The first time Poppy sees Gran, she's standing behind the desk at the Savona Hotel. She's twenty years old, five feet two, her fair hair swept up and pinned at the back. She holds herself erect and turns on him a direct gaze from steady eyes. Poppy is working for the Canadian Northern Railway and has stopped for the night in Savona. He's struck by her poise and self-contained manner. Her face in repose is serene and unruffled, quietly alluring. Poppy can't return to Savona for several weeks but he knows he's going to see Frances Ferguson again.

Poppy proposes to Gran a few months later. They are married on August 19, 1914, a month after the assassination of Archduke Franz Ferdinand that sparked the start of the First World War. Gran is two days short of twenty-one, Poppy twenty-seven. Gran settles into the life of a new bride, first in Kamloops in the interior of British Columbia, then in the Kitsilano neighbourhood of Vancouver. Within the next five years, they have two children: Rowly in 1916 and Frances in 1919. I imagine Gran eager to start her own family, filling the void created by her absent family.

Gran and Poppy were similar in some ways: they were both meticulous in their appearance and both set great store in manners. Neither was frivolous or likely to go off on a whim. They were sensible, reliable, upright citizens. Gran was straight-backed, a small woman with a steadfast demeanour. Her hair turned white when she was in her late twenties, after a car accident in which a friend was killed. She never drove a car after that.

I recognize now that her life was not always easy with Poppy. If everything were not perfect—dinner not ready at six on the dot, the silver tarnished, flowers wilting in the vase—Poppy would voice his complaint. As Rowly matured, his presence sparked ongoing tension in the household. I wonder when Gran recognized that Rowly was different? It must have been obvious from when he was quite young. In photos, the way he stands, the haughty tilt of his head, hands folded precisely in his lap, telegraph his difference.

Some time ago, Mother told me that Gran left Poppy once because of Rowly. When Poppy decided that Rowly should spend the winter with Charlie Lindeman, Gran disagreed and begged Poppy to reconsider.

"It will be too hard on Rowly," she said. "He will hate us if you make him do this."

But Poppy was determined. He packed Rowly's bags and took him to Vedder Crossing to meet up with Charlie in early November. When Poppy returned, Gran was gone. She left a letter for her daughter.

Dear Frances,

You are going to become a boarder at Crofton House School for the rest of this term. I have spoken with the headmistress, and she expects you. I have packed your suitcase with everything that you need. It is sitting at the bottom of the stairs at the front door.

I am going to stay with my sister Rita for a few months. I will write you every week. The headmistress knows how to get in touch with me. If you need anything, ask her to contact me.

Love Mother.

Gran did not leave a letter for Poppy. He returned to find the house empty. His daughter was boarding at her school. His wife was gone, and he knew why. He went to work as usual, he visited Frances at school, and he waited. Two months later, just before Christmas, at the end of Frances's school term, Gran returned. She phoned Poppy at work and asked him to go and retrieve Frances from school. When they arrived home, she said nothing. Poppy understood. He would not send Rowly away again. He would stop trying to change him. When Rowly arrived home in April, Poppy avoided Rowly, but he did not oppose him. Gran redoubled her efforts to connect with her son. She was determined to help him find a path in life that would make him happy.

The author's mother and uncle in Vancouver, circa 1924. Courtesy of Webb family.

In so many ways, Gran was the glue that held the family together. Poppy was the master and commander at the lake but the Gran I knew as a child was a key influence on all aspects of my life.

What did Gran think of the lake? Did Poppy's dream become her dream? I don't think so. She was happy if Poppy was happy and Poppy was never happier than when he was at the lake. She was in her element in the kitchen, her domain at home and at Cupola Lodge. But overall, I'd say that her connection was one of placid acceptance. She was not in the thrall of the lake like Poppy was, like I am.

Chris and I sit on the wooden benches on either side of the kitchen table. Its rough, hand-cut boards, shiny with layers of varnish, reflect the sun beaming through the window. Gran looks at us from her chair at the end of the table, across the debris of our lunch.

"Maybe I'll bake some pies today," Gran says. "Can you two pick some berries?"

Chris and I exchange grins. "Yes," we chorus in unison.

We jump up from the table and go out to the screened-in cooler on the porch, where we pick up empty lard tins with wire handles, just the right size for us to carry. Around the edges of the cabin clearing, we find bushes of huckleberries, a small rosy berry, perfectly round. We pop a couple in our mouths and add the rest to our tins. They are tart, not tempting to eat. At nine, I am an old hand in finding berry bushes. We tramp through the woods up the water-line trail to Cupola Creek. On the way, we stay on the lookout for the blackberries we often come across in clearings in the forest. These add the real flavour to the pies: the huckleberries are the ballast, the blackberries the zing. We pick blackberries off the vine, still warm from the sun. Irresistible! We fill our tins and return to the cabin our mouths smeared with telltale purple.

"We have lots of berries, Gran," Chris says.

"And I see you've eaten a few too!"

Gran puts on her apron, with its flowered pattern and frill along the hem, her hair a white aura around her smiling face. Her glasses are firmly in place on her nose. She has a look of concentration but also patience with our excitement.

Her pastry table, in the centre of the kitchen, is painted pale yellow, with two deep metal drawers, one for flour and one for sugar. Above these are narrow drawers that contain the measuring spoons, rolling pin, fluted cutters, hand blender, sifter, pastry brush, eggbeater, cotton pastry cloth and rolling pin cover.

She sets out her cream-coloured porcelain bowl with a pale yellow border at the rim. She blends the lard with the flour, adding water bit by bit until she has the right consistency. She lays out the pastry cloth and puts the cover over the rolling pin and sprinkles flour over both.

She describes each step: how the pastry has to be rolled to a certain thickness and placed gently into the floured pie plate. Then the berries go in, the pale red of the huckleberry contrasting with the dark blue-black of the blackberry, along with a few spoonfuls of sugar and a squeeze of lemon. The final pastry layer is laid carefully on top, edges pinched to make a ridged crust. She takes a fork in her hand.

"I prick the top to let off steam."

Oven mitts in hand, she places the pie on the middle rack of the oven. Over the next forty minutes, Gran tends the wood stove to maintain a constant temperature.

The wait begins. After about half an hour, the crisp buttery smell of pastry browning and the tangy aroma of bubbling berries permeate the cabin.

We watch as the pie comes out of the oven, perfectly browned on top. It has to "set" and we wait in anticipation until the main course of dinner is finished and dessert is finally served. The first bite of the flaky pastry and juicy berries is worth the wait.

"So good, Gran," Chris mumbles, mouth full of pie.

"You pick the berries, and I'll bake the pie," says Gran, with a twinkle in her eyes.

One thing that Gran can't abide is the smell of fish around the cabin. She insists that fish must be cleaned as far away from the cabin as possible.

Poppy and I return from an early-morning fishing trip. I step out of the canoe as soon as it catches the sand and pull it up onto the beach. I pull out the burlap gunnysack Gran has sewn to make a fishing bag. The bag is slimy, blood oozing at the bottom and heavy. We take the fish out and lay them side-by-side on a cedar board Poppy uses for cleaning.

"Yours is a beauty, Shelley," he says.

I can see that his are bigger but, even so, I bask in his approval. The grating sound as he cuts off their heads with his knife makes me shiver. He makes a single slit up the belly and pulls out the guts. He pokes around to see what the fish has been eating. This helps him choose the bait he'll use next time. I sluice the fish in the water, bits of blood and guts floating around my hands. Poppy cleans his knife and washes out the burlap bag. He makes sure everything is perfectly cleaned.

"We don't want any fish smell when we give these to Gran," he says.

We're now ready to present the fish to Gran. Gran does not fish and she does not clean the fish. But she cooks fish, to perfection.

Gran coats the fish in flour, puts a pat of butter in the hot cast iron pan and lays the fish into the sizzling butter. She turns the fish over to brown both sides and, once its skin is crisp, she places it on my plate. The flesh is pale pink, firm but flaking with pressure from my fork. Delicate and delicious.

Gran generally goes along with decisions made by others. She doesn't impose her ideas. Except once, when she decides we should have chickens at the lake.

"Fresh eggs will be a treat."

I am seven the summer that a crate with two chickens, one red and one white, joins us in the Cessna propeller float plane. Rhoda and Blanche are housed in a wood-and-wire enclosure beside the fish house. Chris, only five at the time, takes turns with me checking for eggs. On the third day, I run back with three warm eggs in my hands.

The weeks go by. We gather eggs each day and everyone is happy with Rhoda and Blanche. Two days before we are scheduled to leave, it suddenly dawns on Chris and me that something is going to happen to our summer companions. They can't stay here alone and they can't come back with us to the city.

"Don't kill them," we plead. Dad says he will look after everything and we won't have to see a thing. That night, a savoury smell permeates the cabin. We sit down to eat dinner.

"Is this Blanche or Rhoda?" Chris asks.

He starts to cry. I fight back tears. Neither of us can touch a bite. We stare down at our hands, anywhere but at the chicken on our plate. Silence settles on the table. Gran gets up and quietly brings us bread and jam and takes away our plates.

Gran never brings chickens to the cabin again.

I keep a journal of stories of the lake that I want to remember. Every now and again, when the grandchildren are at the cabin and we're sitting around the table after dinner, I bring it out. Last night I told them the Blanche and Rhoda story.

"We should get some chickens," says Fin. "We could build some pens for them behind the woodshed."

"But I wouldn't want to eat them," says Nora.

"Me neither," says Will. "Can't we please just buy our chickens at the store?"

It's the end of July and both Kevin and Michael's families are at the cabin for the August 1st weekend. The heat of summer is interrupted by several days of pounding rain. Then, as suddenly as it started, the deluge stops. The sun is out in full force. The water level recedes. Branches and logs bob in the lake. Floating debris washes up on our beach. I wave to Chris and his wife, Janet, as they head out for a morning paddle on their stand-up boards.

I enlist Fin's help to push the logs back into the water so they can float towards the river outlet. A couple of them are too big for us to move. Fin fetches the peavey from the shed. Our peavey is an old tool of Poppy's. It has a wooden handle with a metal spike on the end and an arm with a metal hook. Fin rams the spike into the log, but can't get enough leverage to roll the log over. I join him on the peavey and together we are able to move the log to the water's edge.

"Gran always wanted the beach cleared off," I tell Fin. "If branches littered the beach, she would throw them back into the lake. Even small logs,

The author's grandparents and mother seated on a log at Chilliwack Lake, circa 1935. Courtesy of Webb family.

she would push into the water. If Poppy saw her, he would come down to help her. But mostly, she liked to complete the task on her own."

I told Fin that when I was a girl, our unvarying routine was that everyone rested in the afternoon. Gran and Poppy slept in the cabin. Dad sat in a chair under the birch tree with his book. I lay on the grass nearby with a book, lazy with the heat.

"I'd like that, reading on the grass," Fin says.

"You know, Fin, you come from a long line of avid readers, starting with your great-grandfather."

After her nap, Gran would appear at the door of the cabin and walk down the steps from the porch. I'd hear Dad say, "Oh, oh. There goes Gran, headed for that log on the beach."

We'd both get up from our books, Dad with a quiet sigh, and off we'd go to help Gran.

"So, here we are, Fin. Now you're helping your grandmother move logs off the beach, too."

While Fin and I move logs, the rest of the family is at morning chores. Roger chops kindling outside the wood shed. Will places the sticks into boxes. Fiona drives the John Deere tractor. Nora and Bronwen pile fallen branches into the trailer to take to the burn pile. Kevin digs a fire pit on the beach, placing logs in a circle for an evening campfire and sing-along. Michael cleans the boat. Patrick sets out the sprinklers. Molly and Joanna make sandwiches for a picnic.

We gather at eleven on the dock. Kids in bathing suits carry sweatshirts and towels. Molly and Joanna tote the hampers.

"Okay. Is everyone ready to go?" asks Michael.

"Don't forget the dogs," says Fiona.

Patrick steps into the boat and takes the captain's chair. Everyone settles into the benches in the bow or the stern. Patrick pushes the throttle. The wind is in our hair, the water smooth as glass. I love speeding across the lake, this feeling of freedom.

"Let's stop at Gran's beach," says Kevin.

On one of Gran's last picnics, we stopped at a small beach halfway down the west shore, in a protected cove with a crescent of sand. Gran pronounced it the best picnic ever. From then on, that sickle of sand was Gran's beach.

"Good idea, let's have our picnic there," I say.

At the end of the day, drowsy from the sun and wind, we finish up dinner and the dishes.

"Can we play a dice game tonight?" asks Fiona. "That one about the ship and captain?"

"Okay," says Patrick.

"I'll get the games blanket," says Roger.

"I'll get the pennies," says Will. He climbs onto a chair to reach the top of the bookcase and knocks over the bowl of pennies. He looks dismayed.

"Don't worry, Scooch. I'll help you," says Nora.

"You used to play dice games when you were a girl, right, Rani?" asks Bronwen.

"Yes, but you know what, we used nickels instead of pennies."

Dice games were the evening's entertainment with Gran and Poppy.

After dinner and dishes, Dad drapes a cloth on the dining room table. Gran and Poppy sit at each end of the table and my parents and Chris and I sit along the side benches. Chris places the leather cup with dice and the copper bowl filled with nickels in front of Poppy.

Snake Eyes is our first game. Two dice are placed in the leather shaker cup. Everyone antes in a nickel, and the game ensues. We keep track of the total count of each round. The highest total, once you get a count of over one hundred, is the winner. The hardest part is keeping track of your score and adding the dice quickly in your head. The dreaded pair of ones or "snake eyes" means you go back to zero and ante in another nickel. Whenever any of us has to ante in, we all chime in with the song:

Put another nickel in
In the nickelodeon
All I want is lovin' you
And music, music, music

The next is *Ship, Captain and Crew*. This uses five dice, and is easier and faster. On every turn, you get three rolls to get first a six (ship), a five (captain), and a four (mate). The player who gets the highest count on the last two dice (crew) wins.

Gran likes dice games and usually wins. When I'm older, I realize that every now and then she lets Chris win. He gets so excited. He jumps up and dances around the table. Gran looks over at me and gives me a smile.

When Poppy dies, Gran is seventy-nine. She no longer wants to stay alone in her cabin. She suggests that Chris and I share the use of Cupola Lodge.

It is the summer of 1975. Patrick and I have driven out from Calgary where we are living; the boys are three and one. Living in my grandparents' old cabin is like a homecoming for me. We put extra beds in the bedroom. We bunk in together. From then on until we build our own cabin, we spend our summer holidays at Cupola Lodge. Gran comes to visit when we are at the cabin but stays with my Mother and Dad at the Sandpiper.

Over the next ten years, Gran comes to visit us regularly in Calgary, flying out on her own, coming for a week in the spring and fall. We are lucky. After Poppy dies, we have another eighteen years with Gran. Patrick adores her. He fusses over whether she is warm enough in our cold Calgary climate: brings her a shawl to put over her shoulders. He sits beside her on the couch showing her the recent photo albums while I make dinner. I smile as I hear her soft laugh in the next room. Patrick will have told her a joke or teased her about something. Kevin and Michael join them in the living room as soon as they're home from school and the laughter becomes more raucous. When I come in to call them for dinner, Gran's face is wreathed in smiles.

I become a volunteer with the Young Women's Christian Association in Calgary, joining the board of directors, eventually taking on a term of president for two years. I find out later that Gran was also on the board of a YWCA—the Vancouver branch—when she was a young mother. She rarely talked about herself so much about Gran is unknown. She was always the listener. When I was a child, I was told to "be seen and not heard." I wasn't encouraged to ask questions, certainly not personal ones. The reticence of

Gran and Poppy and my mother too, has motivated me to be more open with my grandchildren. Their generation is inquisitive. They challenge the opinions of adults. I like that my grandchildren ask me difficult questions and expect an honest answer. My research this summer is exposing some uncomfortable history. My grandfather's treatment of Rowly. Gran's sacrifice of her art. My mother's mercurial personality. I don't want to be a keeper of secrets. Bring them out into the open, that's my philosophy. And so, I share these stories with my grandchildren. If we don't talk, we'll never learn.

Gran's last visit to the lake was when she was ninety-two. It was 1985. We had just moved to Vancouver where I accepted a position as an articling student at Bull, Housser and Tupper, a Vancouver law firm. We drove her out for the August twenty-first weekend, the last of Gran's birthdays to be celebrated at the lake. Chris and I and our families sang her favourite songs, bringing back "Five Foot Two, Eyes of Blue" and another oldie, "Sergeant Major" that we sang in a round.

> Around the corner, and under a tree,
> A Sergeant Major made love to me.
> He kissed me once—he kissed me twice,
> It wasn't the proper thing to do,
> But gosh, it was awfully nice!

Gran got such a kick out of these somewhat risqué lyrics! I think she knew it was her last time. She was treasuring every minute. She never saw the cabin that Patrick and I planned to build, although we told her about it. She died six years later in a nursing home in Vancouver.

I am inside the cabin, listening to Kevin outside on the porch talking about Gran to his children. He was eighteen and Michael was sixteen when, at ninety-seven, she died quietly in her sleep.

"I was lucky to know my great-Gran well. She would come for dinner every week. I would arrive home from school and Gran would be sitting in the den, waiting for us. Your uncle Mikey or I would get out the latest photo album and sit beside her on the couch going over the photos with her. She loved the stories that went with the pictures. Her hands would be folded in her lap, and she'd twiddle her fingers round in a circle and smile, just happy to be with us.

"When Gran died," Kevin continued, "we had a memorial service for her at the lake. We sat in a circle and said prayers. Chris played the guitar while we sang a hymn, and we each told our favourite stories of Gran."

I remember. I started. "When I think of Gran I think of Purdy's chocolates. Did you know that Gran started the tradition of Purdy's chocolates? Gran not only brought a box of Purdy's to the lake but she also brought them each time she came to our house for dinner."

Michael was next. "One day, Kevin and I were out exploring in the woods. I stepped on a wasp nest and got stung all down my back. I came out of the woods crying. Gran took a hold of me, told me to lie face down on the kitchen table. She made a paste of water and meat tenderizer and plastered it over my back. No more stings. No more swelling."

Patrick had a story too. "Here's another of Gran's remedies. If I got sap on my hands she told me to use butter. Take some butter, rub it on the sap and the sap will come off. It works!"

Bronwen interrupts Kevin's reminiscence. "Rani uses the butter trick."

"When we finished with our stories," Kevin says, "we said a final good-bye and we put Gran's ashes in the flower garden behind her old cabin, Cupola Lodge."

After Gran died, I wanted to commemorate her place at the lake. We called the creek that supplies our water, Cupola Creek. We called the creek that flows into the bay after Otter Cove, Frances Creek, after Gran. In 1998, I applied to the Geographic Names Office of British Columbia to have these two creeks registered with the names we'd given them. My rationale:

> Cupola Creek has been used as a water source for the Webb families since the 1920s. Frances Creek is requested to be named for Frances Webb, born Frances Ellen Ferguson in Savona BC on August 21, 1895, educated at Old Hallows School in Yale BC, then at St. Ann's Academy in Victoria, married Christopher Everest Webb in 1914, children Rowland and Frances. Frances Webb first arrived at Chilliwack Lake in 1923, and lived for the summers in a one-room log cabin on the shores of the lake from 1924.

My request was approved a year later, and so we now have Frances Creek, as well as Mount Webb, both named for our family. It seems appropriate somehow, having a mountain named for my grandfather and running water for my grandmother. Poppy was an imposing presence, hard and implacable. Gran was flexible, someone who always went "with the flow."

The name Frances lives on in the next generation. Actually, the name connects our two families, Patrick's and mine. My grandmother and mother's first name was Frances. My middle name is Frances. Patrick's mother's maiden name was Francis. Patrick's middle name is Francis. Our son Michael's middle name is Francis. And continuing the tradition, our granddaughter, born in 2001, is Fiona Frances O'Callaghan.

Gran was one of those quiet presences that seep into you slowly. She was never at the forefront, neither the teller of tales nor the centre of attention. She encouraged others to assume that role. Poppy was proud of us, especially when we rose to meet the bar he set, but I did not feel the same sense of comfort and acceptance that I felt with Gran. With her, I could be who I was. I never felt her passing judgment. Gran was loved and admired by everyone who met her. Not that she was a milksop or a doormat. She had backbone. But her strength was subtle. As I found out, she could walk out on Poppy if need be.

As Gran got older, I felt protective of her, especially with my mother. Gran would occasionally come to the lake in her late eighties and stay at Sandpiper with Mother and Dad. Patrick and I now stayed at Cupola Lodge with the boys. I would stop in and visit with Gran every morning.

"Hi, Gran. Hi, Mum," I say, walking in the door of the Sandpiper.

Gran is in the kitchen sweeping. I see that my mother is putting out slices of bread for sandwiches.

She lashes out at Gran. "Mother, you are in the way. Just put the broom down and get out of the kitchen. Can't you see I'm getting ready for lunch?"

How can she speak to Gran like that? I hurry into the kitchen, walk right past Mother and come up to Gran, giving her a hug.

"Come on, Gran," I say, "let's go sit outside on the porch. It's nice in the sun. The boys will be here any minute. They want to show you their new soccer ball."

As I get older, Gran is the person I most want to emulate. I like the things that connect us. We are both small in stature, understated, not flashy. Listeners. She had qualities that I value: she was non-judgmental, thoughtful, selfless and caring. I recognize that I am more stubborn and more ambitious than Gran. But she is the star that guides me. She is my life compass.

MOTHER

"How old were you when you first came to the lake?" Bronwen asks.

"Six weeks."

"So, I'm the youngest," she says. "I came at three weeks."

Bronwen has found her place in the history of the lake.

Bronwen and I visit my mother at Crofton Manor right after the August 1st weekend. The heat of summer is full-blown. I suggest that we go out to the courtyard to see the geranium bed she planted with salmon pink flowers. I open the door and Bronwen pushes Mother's wheelchair into the courtyard. A hot breeze blasts through the space. Heat radiates off the pavement. Mother recoils from the glare and waves her arms in front of her eyes.

"We should go in, Rani," says Bronwen. "It's too bright."

"Good idea, Bron."

We retreat to the dimness of Mother's room, curtains closed, lights off. She sighs with relief. Mother is alert today. I remind her that sixty-six years ago I came to the lake for the first time.

"You were just a baby. I brought you in on the float plane," she says, pointing to a photo on the dresser.

Bronwen picks up the framed black and white. At thirteen, Bronwen is deeply curious about how things used to be done. Mother stands on the pontoon of the de Havilland Beaver single engine float plane. Wind whips her hair off her high forehead. She shows me off to the camera, a small bundle wrapped in a blanket, only a shock of dark hair and peeping nose in view. Mother has high cheekbones and brown eyes set off by thick brows: striking, both in looks and temperament. She is wearing plaid slacks, a white blouse and cream sweater. Her slender figure belies the fact that she gave birth only six weeks before. In another photo, she rinses diapers in the washtub perched beside a wooden Pacific Milk box. Clean diapers are hanging from wooden clothespins clamped to a clothesline strung between the porch posts. Her white shirt is rolled up to the elbows. Her hair falls damply over her eyes.

"I put you to sleep on the porch in a dresser drawer," she says now.

Bronwen looks up at me, surprise in her eyes.

"Can you imagine that, Bron," I say. "A one-room cabin with a six-week old baby, nothing but cold water running from a hose and an outhouse in the woods. No way to call for help."

"Wow!" says Bronwen. "I guess no Pampers."

Bronwen has recently completed her babysitting course. When she came back from looking after a neighbour's baby, her only comment was, "Changing diapers is a lot of work!"

"No, cloth diapers, washed by hand."

"Yikes, she'd have to use a diaper pin!" says Bron. "I couldn't do that."

Mother had been coming to the lake since she was four years old, but as comfortable as she was with life in the wilderness, bringing a newborn must have been daunting. I think about the time Patrick and I were looking after Fiona at the cabin and she fell terribly sick with asthma. She was only nine months old and I got so scared that I drove her to the Chilliwack Hospital in the middle of the night. I could get help in less than an hour. What would Mother have done in such a circumstance?

We show Mother some recent photos, one of Bronwen's dog, Murphy, running on the beach and another of Bronwen and Fiona paddling the canoe.

"Do you remember the old dugout canoe?" asks Bronwen.

"Yes, of course," Mother answers sharply. I smile. She hasn't been to the lake in six years, but she still knows everything.

The author's mother fishing in a dugout canoe in front of the floatplane that has brought the family in for their three-week stay. Barney, her water spaniel, is standing in the lake nearby, circa 1935. Courtesy of Webb family.

Patrick and I walk into Crofton Manor on an early September morning in 2009. We knock on Mother's door. It's funny. When I am with her, I call her "Mum" even though she always refers to herself as "Mother." Her birthday cards are signed "Love, Mother." A sign of our uneasy relationship, I suppose. Even at ninety, she is ready for us, her white leather overnight case waiting by the door.

"Hi, Mum," I say. "Are you looking forward to coming to the lake?"

"Yes, I can't wait."

I help her with her navy rain jacket and lock her door behind us. Mother shuffles along the hallway, pushing her walker. Patrick holds the front door open and leads her out to the car. I guide her into the front seat. Patrick leans over to click her seatbelt in place. We start the car and wend our way through the traffic in Kerrisdale out to Highway One. By the time we hit the highway, Mother is asleep, head slumped against the side window. I lift her head up and slide a pillow underneath it. Two hours pass by.

"We're here, Mum," I say as we drive down our access road and round the corner to the cabin.

"Let's get you settled," I say. Patrick comes around and helps Mother out of the car and up the few steps to the front door of our cabin. He gets the walker out of the Jeep and brings it in the house for her. She immediately pushes it to the front window with its magnificent view down the lake. She smiles.

"It's just as I remember it."

I make her favourite lunch, a tomato and cucumber sandwich, with mayonnaise, of course, and a small glass of milk. After lunch, I walk her across the porch to the Brook, the small cabin where Patrick and I sleep. She'll stay here so she doesn't have to climb upstairs to a bedroom. I pull the covers down and she gets into bed for her afternoon nap.

When I come over an hour later to check on her, she is sitting on the edge of the bed. She looks upset.

"What's wrong, Mum?" I ask.

"I've lost my wallet. I must go back."

Her chest is heaving, her face is flushed. She looks agitated.

"I don't feel well. I feel dizzy."

"It's okay, Mum, just lie back on the bed. I'll look for your wallet."

I call out for Patrick and he comes running.

"I think we should drive her to the hospital," I say.

Mother lies on the bed while we get ready to go. She keeps repeating that she wants to go home. I am alarmed. Is she having a heart attack? I ask her whether she has any pain. She says no. I find the wallet; it's in her suitcase.

Patrick brings the car around. We quickly return everything to the car and lock up the house. He guides Mother back to the driveway, sets her up in the front seat and we drive off. Once we are back on the Chilliwack Lake Road, she falls asleep again. So maybe she is fine, just got panicky with being in an unfamiliar setting. She has lived at Crofton Manor for three years. Although she resisted moving to an assisted living facility at first, it's what is familiar to her now.

We take Mother into the Emergency at the Chilliwack Hospital. The doctor takes her blood pressure. He confirms that she has anxiety. He gives her some medication and says we can keep on going to Vancouver. What a relief. Mother continues her refrain that she wants to go home.

"Yes, we'll take you home, Mum."

We have an uneventful drive the rest of the way to Vancouver, arriving just before dinner. Mother is visibly relieved once we usher her in the door at Crofton Manor.

"Did you have a nice visit to Chilliwack Lake?" asks the receptionist at the front desk.

"Yes," says Mother. "But I didn't want to stay. I'll go in to dinner now."

"Goodbye, Mum," I say. "Are you sure you're okay?"

"Yes, fine."

Patrick walks her into the dining room and ushers her into her chair at the table. I take her suitcase up to her room and put her things away. I look in the dining room on the way out. She is sitting at her usual place, eating her dinner as if nothing had happened. I go and give her a kiss goodbye. Patrick has let the nurse know what happened during the day and handed over the medication prescribed by the doctor in Chilliwack. The nurse says she'll keep an eye on Mother overnight.

That was the last time my mother was at the lake. She first came as a four-year-old, perched on the front of her father's saddle. Eighty-six years later, she drove out in a Jeep.

What was Mother's relationship to the lake? I know the facts of her first trip in to Chilliwack Lake, but I wish I knew what she felt as a young girl. It's the end of the first week of August and Bronwen, Fin, Will and I are expecting Kevin and Joanna after their week of work. The kids and I are playing a game of Ticket to Ride when we hear the beep of a car's horn and look out to see Kevin's car coming around the bend. The kids jump up, excited to see their mum and dad. Murphy, their golden doodle, is whining at the back door. He knows they're here, too. We go out on the back porch to greet them. Kevin embraces me with a big smile on his face.

"Ma, do you remember that summer I tried to get Noni to talk on tape about her early days? It was maybe fifteen years ago."

"Yes, but you said that you didn't get much."

"Right, but I came across two tapes when I was cleaning out my desk last night. I don't know if anything's on them but here they are. Oh, and here's a tape player you can use."

"How amazing!" I say. "I thought it was a complete failure."

"Maybe it was. I haven't had a chance to listen to them. I thought I'd just bring them straight to you."

"I'll listen to them later this afternoon."

"Don't get your hopes up. There may be nothing."

We have our lunch on the deck. Joanna asks if I want to join them on a boat trip.

"No thanks. I think I'll see what's on the tapes Kevin gave me."

I head up to the attic of the Brook where my writing desk sits. I slide the tape into the recorder and press play. Nothing but static. Then, over the static, I hear Kevin's voice, very faint. I put on earphones and turn up the volume as high as it will go. I can just make out what he is saying. Then I hear Mother's voice. I strain to hear.

"I first came to the lake when I was four. Father had to wake up Ed Bell when we got to his farm. It took them hours to get the packhorses ready. Mother was not an outdoors person. She was feminine and wanted to walk. She didn't like to ride the horses. Mother made picnic lunches. We always had a chocolate bar for lunch. It was very important. I remember shafts of light shining through the forest. As a girl, I liked that part. The last part of the trip was the hill, then we were at Charlie's cabin. I started to run as soon as I saw Charlie's. I ran straight into the lake, clothes and all. We walked along the beach and got the green and black boat. Then Father went back and forth to bring everything to the cabin."

Mother's voice then gets drowned out by a passing boat. Kevin comments on how loud the motorboats are. The thread of her first trip is lost. But it's enough. I can tell my mother loved the lake right from the start.

I can see the tension between Mother and Poppy in a photograph in our cabin. She is twelve, standing beside her father. They are dressed like twins in matching long-sleeved shirts, striped V-neck wool sweaters gathered into belted wool knickerbockers tucked into long socks. Leather hiking boots and checked bandanas. They are on a hiking trip in the mountains. Both stand wide-legged, hands on hips, resolute and determined. Both stare unsmiling into the camera.

Mother told me that she eschewed everything feminine. She rebelled against Gran's urge to dress her in pink and petticoats. She was daring and fearless. She was the boy, and she loved it.

My mother, Frances, is twelve. She tells Gran and Poppy at breakfast that she is going to fish on the river today.

"Frances," says Gran, "I don't want you to go on your own. You're too young. Chris, can you go with her?"

"I'll be fine," Frances says. "I know every inch of the river."

"She can go," says Poppy. "She knows how to look after herself."

Gran is unhappy to be overruled. Frances collects her bag, net and fly rod from their hooks on the cabin wall. She calls to her dog, a water spaniel named Barney, and heads across the beach to the rowboat. She's not allowed to take the dugout canoe to the river. It's too unwieldy in the swift current. The rowboat is more manageable. Gran brings a bag down to the boat.

"Here's some lunch," Gran says. "Please be careful and be home before dark."

"I will."

Frances waves goodbye and starts rowing. Barney curls up at her feet and looks up at her, his tail thumping the bottom of the boat. She reaches the river in twenty minutes, pulls into shore, and ties up the boat.

Frances puts her lunch and net in the fishing bag, slings it over her shoulder and picks up her rod. She whistles to Barney and they set off down the trail along the river. After ten minutes of walking, she stops to fish. She places her bag on the rocks and shimmies out on a log to a large rock where she can cast into a deep pool in the river. The water is high. Rapids swirl under the logjams. She has no luck at the first pool and continues walking downriver. She stops at the next pool. Barney is behind her, running to catch up. She glances up just as Barney loses his footing and falls into the river. She throws down her rod and jumps in after him. She manages to grab Barney's neck, then they both are swept downriver a ways before she manages to pull him to shore. Barney is none the worse for wear. Frances is sopping wet and shaking. It was a close call. She knows how lucky she was. She peels off her clothes and lays them out on the rocks to dry in the sun. She eats some lunch, giving the leftovers to Barney. When her clothes are dry, she puts them back on. She keeps on fishing. She gets home just as the sun is setting. When she returns, she has a bag full of trout and a smile on her face.

I don't know whether she tells Gran and Poppy what happened with Barney. I don't know if she was punished. But it was her favourite story and she repeated it often. By the time she told me, she was clearly the heroine of the story.

As she entered her mid-teens, Frances became even more independent. She wanted to stay at the lake on her own with friends. Poppy said she could bring her friends but not on her own. She invited the Amici Club: a group of girls from Crofton House School in Vancouver, where Mother was a student from grade nine to twelve, where she was sent to board when Gran left Poppy for those crucial months. She wanted to show her friends how comfortable she was in the wilderness. Her friends knew only city life; she knew so much more.

The author's mother and grandfather holding rainbow trout caught in Chilliwack Lake, circa 1935. Fishing morning and evening provided the evening meal each day. Courtesy of Webb family.

The four teenage girls come by car from Vancouver to the rail terminus in the Chilliwack Valley. The girls climb on the speeder railway car and grip the side handles on the car for the twenty-four kilometres to the end of the rail line. Frances meets them at the end of the line and the girls ride the final six kilometres on horseback. At Charlie's cabin, Frances unloads the girls' bags and heaps them into the rowboat so she can row them down the beach to her family's cabin.

Everything is foreign to these girls, as if in those few hours they have travelled to another country, or another time. The bathtub is a metal washtub, pale grey with a faded red line, just big enough for a small person to fit in. It is filled with hot water from the large aluminum kettle resting on the edge of the stove. A wood screen is set up around the tub for privacy. The washing "machine" is a washboard with its metal ridges set in a rectangular wooden frame. A clothesline is stretched from the back porch to the woodshed, a cloth bag of wooden clothespins hanging at the ready. The bathroom is a two-hole outhouse set in the woods down a path from the cabin. The bedroom is the porch where Frances has set up four extra folding canvas army cots with wooden frames.

Frances showed her friends how to catch and clean fish, how to set a fire in the stove, cut kindling and trim the coal oil lamps. I suspect that she was displaying her talents not only to the girls but also to Poppy who could see for himself how well she managed. She was independent and resourceful. Competent. Always right. The lake provided the stage on which she could trot out her accomplishments. As it did for Poppy. As it does, I realize, for me.

In the mid-1930s, friends of the author's mother travelled the length of the Chilliwack River valley on a speeder car to the end of the logging railway. Here they saddled up packhorses and rode the remaining six kilometres to Chilliwack Lake, a journey that took them most of the day. Courtesy of Webb family.

My mother's life had the patina of perfection. She excelled at school, she was popular, and she emerged in her teens an arresting beauty. She was ambitious. She completed a Master of Social Work at the University of British Columbia at a time during the Second World War when other girls in her social circle were preparing to be wives and mothers.

She joined the Women's Royal Canadian Naval Service in 1942. She was stationed for the duration of the war in Halifax, providing radio communication services. She experienced the dread of watching waves of planes leave the base, knowing that many friends would not return. Even so, she thrived. Her life had a purpose. She was as good as any man. That was what drove her: to live up to Poppy's expectations. To be the son he wanted in place of Rowly. Was it this quality of driving ambition that attracted my father?

My mother met my father, Gary Bowell, through a mutual friend. Dad came from a working-class background. Raised in Nelson, a small town in the West Kootenay region of southeastern British Columbia, he worked his way through school. He qualified for scholarships, receiving a Bachelor of Arts in history and economics from Queen's University in Kingston, Ontario. Through university, he came home once a year, working his way across Canada as a porter on the Canadian Pacific Railway. He was awarded a Rhodes Scholarship, but war was declared and so instead he joined the Canadian Artillery. After the war, he took his Masters in Business Administration at Harvard University.

It wasn't the way things were done then, but Mother married Dad within a few months of meeting him. A surprising coincidence with the short time Gran and Poppy knew each other before they married. Gran and Poppy didn't actively disapprove of Dad, but I suspect it was a long road to acceptance. Dad was a peacemaker, bending to Mother's mercurial moods. He was scholarly and mild-mannered, with a gentle smile and curly blond hair that thinned early, leaving him with a white fringe like a monk's tonsure on his bald head.

Dad spent his whole career in the forestry industry, starting in the White Pine Sawmill with MacMillan Bloedel. He rose in the ranks to become executive vice-president and left when appointed the president of Weldwood of Canada. When Dad started in on any topic related to forestry, I settled in, knowing that Dad could keep talking for hours. He loved nothing better than wandering the woods, pointing out to Chris and me distinguishing features: bark, needles, limb shapes, the crown of the tree.

A reader and philosopher, he spent his last ten years researching and publishing a book: *Grass Roots Peacemakers: The Free World Wake-up Call.*

I put in the second tape Kevin brought me. I hear Dad's voice through the static. He is describing his first visit to the lake.

"When was my first visit? Let me think. Shelley was just born and we left her in Vancouver. It must have been early July 1948. We thought of it as an adventure. It was our closest friend Gordie's birthday and Frances wanted to take Gordie and his wife Ruth to the lake. And me, I had never been there. We stayed in a motel in Vedder Crossing, then drove in the morning to Bowdenville. This farm was owned at the time by Gordon Wells. We arrived about eight in the morning. Nothing was ready. Getting the horses saddled up and packs on: that took about an hour. We had five horses, one for each of us and one for Gordon Wells. We rode until about three in the afternoon when we came to the Chilliwack River crossing. The river was too high for the horses to cross. Frances was determined to get in to the lake. She was not turning back. There was a log fallen across the river. She said we could shimmy across the log and walk in to the lake from there. We took the packs off the horses and put the essentials into our backpacks. Gordon Wells said he would take the supplies back and his brother Ronnie, he was a pilot, would fly them in the next day. Each person chose a few things to carry. I chose a bottle of scotch. Ruth chose a cardboard box. She wouldn't say what it was. It turned out it was a cake mix for Gordie's birthday. Frances got over the log no problem. Ruth got part way along and said she just couldn't do it. Gordie and I got Ruth in between us on the log. She shimmied across with us encouraging her. We said, "Don't look down at the water. Just look ahead." She made it. It took us three hours to walk in to the lake. The first job was to pull the boats out of the cabin. Then we had to go up into the attic and get the mattresses from the tin box. We pulled them down the narrow stairs from the attic and put them on the wooden bed frames on the porch. Then we brought down the blankets, sheets and pillows. It took us at least two hours to set up the beds. We had a cold supper, I don't remember what. But we finished up with a glass of scotch, totally exhausted. That was my initiation to the lake. A great adventure."

Dad loved the place right from the start. It gave him the space to relax, to lose himself in his beloved books. I am sure he was relieved when he and Mother were able to build their own cabin, the Sandpiper. And like Poppy, he had some creative solutions to challenges.

By a quirk of fate, bats took up residency in the eaves behind the chimney of the fireplace of the Sandpiper, too. Luckily, they were on the outside of the cabin.

The squeaking of the bats starts up at dusk and continues through the night. Dad puts up a "bat house" at the edge of the clearing near the cabin. The bats don't budge. I have a vision of hordes of bats breaking through the cracks in the logs and taking over the living room. Gran once told me that bats can get tangled in your hair. If I see a bat in the cabin, I dive for cover, hands over my head, protecting my hair.

Over the years, Dad tries all sorts of methods to get rid of the bats. None of them work. Years later, when Kevin and Michael are ten and eight and we are spending the summer at the Sandpiper, Dad comes up with a new plan. Waiting until dusk, he rigs up netting from the roof eaves across the end of the cabin with the fireplace. He positions Patrick and I, Kevin and Michael on the porch outside the netting and hands us various bludgeoning instruments—baseball bats, tennis racquets, brooms. He brings his Chevrolet station wagon around to the edge of the porch and hooks up a long hose to the car's exhaust pipe. He turns the car on and puts it into neutral. Then he positions the ladder against the roof close to the chimney. Dad climbs the ladder and holds the hose to the cracks between the chimney and the logs. Car exhaust spews from the hose.

"We're going to smoke them out," says Dad.

The posse stands expectantly. The squeaking gets louder.

"Look! Here they come!" shouts Michael.

A flurry of drowsy bats fly out, getting caught in the netting.

"Quick, boys!" Dad shouts.

We start whacking away at the bats in the netting. I stop, horrified at what I am doing. I can't hit any more of the creatures. We look sheepishly at each other.

"Okay, that's enough," Dad says awkwardly.

I think he wishes he never thought of this idea.

How do I feel about how we treat animals here at the lake? Some, I feel justified in killing, and quite cold-heartedly. Others I want to protect. Why is this? I was taught that any animal that came into the cabin was vermin and should be eradicated. Mice and rats fell into this category. I didn't consider whether these creatures were plentiful or endangered.

I look up "mice" online. Our mouse is the Western Harvest Mouse. I am shocked: it is a protected species in British Columbia. And here we are killing them by the dozens without a second thought. Is there another way to keep mice out of the cabin? I need to think about this. And what about bats? I cringe now when I think about killing all those bats with my dad thirty years ago. We don't kill bats now. In fact, we have a resident bat in Creekside that swoops across the living room at night. The girls scream but everyone else ignores it. I now repeat Poppy's advice: bats are good to have around; we should protect them.

Every afternoon, Dad sets a lawn chair in the shade of the birch tree in front of the Sandpiper and sits down with a good book. A devoted reader, he is content for hours. With his white cotton brimmed hat shading his eyes, he sometimes nods off, the book open in his lap.

One afternoon, he is awakened by a noise behind him. He describes it later as a snuffling. About three metres away, a black bear moves towards him.

"Steady girl," Dad says. "Good girl," as if he were encouraging one of the dogs.

The bear edges closer.

"Why did you think the bear was a girl?" I ask later.

"I don't know," Dad says. "It's just what came to me."

The bear hesitates.

"Carry on then," Dad says firmly.

And she does, ambling on to the beach, down to the water's edge, where she leans in to have a drink. All this time, Dad has not moved a muscle. The bear looks up and keeps on walking down the beach. Dad, when later recounting his tale, describes it as peaceful—just two old folks passing by, acknowledging each other's presence.

Dad was dazzled by Mother. I don't think he ever fully believed that she deigned to marry him. She could do no wrong. Mother called the shots and Dad would do anything to make her happy, which was a futile task since my mother was consumed by discontent. Perhaps she was born into the wrong era. She stopped work when I was born but to me she never seemed fulfilled by the role of wife and mother.

Mother's mood swings were severe. Sometimes she was elated, sometimes despondent. Mostly, she was restless. I learned to sense her dark moods and escape, taking Chris with me.

I am nine and Chris, seven. We are living in Port Alberni, a small forest-industry town on Vancouver Island. As manager of the pulp mill, Dad—and by extension, Mother—are important figures in the local social hierarchy. Mother chafes at the narrowness of small town life. She doesn't make friends easily. Chris and I arrive home from school one afternoon in early fall. I can hear shouts and pots crashing in the kitchen. She has been preserving raspberry jam for the past two days. I grab Chris's hand and go back out the door. We retreat to the back of the garden where we are hidden by brambles and bushes. We whisper stories to each other. I wait until I see Dad's car sweeping in the driveway. It's safe to go in the house now.

When I was a teenager, Poppy stopped going to the river with his fly rod. Mother and Dad decide to have a family day of fishing on the river, just the four of us—Mother, Dad, Chris and me.

Dad assembles the fishing gear the night before. I make sandwiches for our lunches. We take the green and black rowboat with the five-horse Johnson motor. We arrive at the river outlet and tie up the boat. The river puts a smile on Mother's face. She likes nothing better than fishing on the river. She hums to herself as she ties her fly to her line. She sets off down the trail along the river's bank. She's heading towards her favourite pools. She is at ease, her stride confident. Dad stays back with Chris and me, happy to give Mother her own space. We meet for lunch and Mother, laughing, shows us her catch. Four gleaming trout, each twelve inches. She is sparkling. Radiant. Ebullient.

But the lake was also where she wanted to exert control. Like Poppy before her. And often the place brought out the worst in her. Her brother, Rowly, was mostly absent from the family while he lived in England. After Poppy died in 1974, he returned and announced to Gran and to his sister that he was gay. Soon after, he brought his partner, Richard, to the lake for a weekend.

Rowly stays with my mother and father at the Sandpiper. Patrick and I and the boys are staying at Cupola Lodge. Mother invites us over. We put Kevin and Michael to bed early in the front bedroom at the Sandpiper and join them for dinner. We have roast beef and Yorkshire pudding. Mother is unpleasant and demanding. She fusses over details. She starts ordering Rowly around.

"Rowland, light the candles."

"Rowland, clear the table."

"Rowland, you and Richard start the dishes."

"No, I am not doing the dishes."

"I cooked the dinner. The least you can do is help with the dishes." Her tone is light but cutting, like a stiletto. I can't believe that she is being so rude.

"We are leaving," Rowly says abruptly. "Richard, get your things."

As he walks out the door, he turns to his sister, "We will never come back."

It is nine o'clock at night. They get in their car and drive away. I am shocked and so is Patrick. We pick up the boys and leave for our cabin, thankful to get away.

My mother never forgave Rowly, and it seems, he never forgave her. I was more sympathetic to him than to her. I was close to Rowly. In my late teens, he would invite me to his apartment for lunch of cucumber and watercress sandwiches. He gave me elaborate gifts: eight antique dining room chairs for my eighteenth birthday, a gold bracelet carved by Haida artist Arthur Adam for my twenty-first birthday. I was saddened to see my mother and uncle's relationship deteriorate. At one time, when Rowly first returned from England, he and Mother were confidants. Mother asked Rowly to help decorate her house. They would return from Maynard's Auction House delighted with their "finds." How did this closeness evaporate?

Through most of my life, I tried to get as far away as possible from Mother. I travelled during summer breaks from university. Patrick and I moved to Zambia right after graduation. We lived in London, Ontario; Toronto; and Calgary; anywhere but Vancouver for nearly twenty years. Mostly, I would see Mother at the lake, and often those visits were fraught with tension.

Once I had children of my own, I thought Mother and I would grow closer. But that didn't happen. She tried to rule how we raised our boys. She just couldn't seem to stop herself from making comments.

"Kevin shouldn't have any more cookies."

"You have to let Michael cry himself to sleep."

After Poppy died, she made it clear to me that she was the one in command at the lake. The matriarch at last. One summer, Patrick and I and the two boys, who were three and one, came and stayed at the Sandpiper for a couple of weeks. Mother and Dad left us to enjoy a few days on our own. The lower limbs of the large birch tree between the cabin and the beach were blocking my view from the kitchen window. I asked Patrick to trim them back so I could see the boys playing on the beach. When Mother and Dad arrived, Mother was furious.

"How dare you take limbs off the tree without asking me first?"

I was not as contrite as she expected. "You know, Mum, all we did was clear a view to the beach. I wanted to see that the boys were safe."

The tension never eased through the rest of our stay.

But I have to admit that Mother's stubbornness and perseverance has benefitted us, too.

It's a hot August weekend and everyone is at Creekside. At the end of the afternoon, a flotilla of boats sets off towards Otter Cove. The cove catches and holds the last rays of the sun and the rocks along the shore provide great jumping spots into the water. Roger and Fin are in the Zodiac with the five-horse motor. This inflatable small boat has been constantly in use all summer, providing the independence the grandkids crave. The rest of us are on paddleboards, stand-up boards and canoes. The late afternoon sun sparkles across the wake of the Zodiac. The two dogs, Dennis and Murphy, run along the shore, not wanting to be left behind. I sit on a paddleboard with Fiona, Bronwen on her own board beside us. We straddle our boards, our legs in the water.

"Where is our property line, Rani?" Fiona asks.

"You see that broken snag on the shore? That's the boundary."

"Our property goes up to the forestry road, right?" Bronwen asks.

"Yes, it does."

"When was the road put in?" asks Fiona.

"Over fifty years ago. My mother, your great-grandmother, fought to keep the road as far from the cabins as possible. The Forest Service wanted to put in a road along the eastern shore of the lake to the south end. They wanted to cut through our property right behind our cabins."

"What!" says Bronwen, outraged, even half a century later. "That'd be awful!"

Mother gets out of the panel van to open the gate to our access road. It is the fall of 1964. She notices orange tape tied to nearby trees. The markers veer off in a line across the family property down to the beach. She is incensed. She strides into the forest and tears the tape off every tree. Dad hesitates, then follows. The taped trees follow the shore just behind Cupola Lodge and Sandpiper and curve into Otter Cove to the eastern shore of the lake.

"This must be the line the forestry road will take," says Dad.

"They can't put the road here!" Mother declares.

Back in Vancouver, Mother phones the Ministry of Forests. The original road to the lake had been completed eight years earlier. The ministry confirms that they plan to put a forestry road down the east side of the lake to the south end. The proposed route will cut right through Cupola Estates property. Mother asks for a meeting with the minister in Victoria. Poppy comes with her. At the meeting, Poppy argues that the road should not be built at all. The high construction cost and expense of ongoing maintenance will be prohibitive. He describes the difficult terrain. The east side is steep and mountainous. The lack of topsoil creates unstable conditions. The proposed route crosses dozens of creeks and rivers. Avalanches are frequent during the winter; the creeks and rivers flood in fall and spring.

The government is determined that the road will go ahead. In early 1965, Mother and Poppy attend a second meeting with the deputy minister to discuss the exact location of the road. The official takes a hard line. The government has the right to put the road wherever it wants. Mother unfolds contour maps to point out an alternative route that won't go directly beside the family's cabins. She is worried about intruders. If the cabins are visible from the road, they will draw unwanted attention. She requests that a surveyor comes to the lake to look at other options. Finally, the deputy minister agrees.

Mother meets the surveyor several weeks later and together they walk the property. Mother keeps trying to push the route further from the cabins and higher up the hillside.

In August, the deputy minister writes, advising that the right of way for the forest road has been located as far up the mountainside as is feasible and, as negotiations have extended for a considerable period of time, this is the ministry's final position. The final route requires a transfer of two acres from Cupola Estates to the Ministry of Forests. In September, the right of way location is published in the British Columbia Gazette.

Neither Mother nor Poppy sees it as a victory.

"I guess it's final," she says. "I am not happy but I pushed them as far as I could."

Her work doesn't end there. The following spring, construction of the forestry road begins. Every morning, she walks up the road, survey plan in hand, to make sure the bulldozers excavating behind our cabins are keeping to the agreed-upon route. Her mantra: Keep a high line behind our cabins and stay above our water intake on Cupola Creek. The contractor does his best.

As the road is pushed ahead, the debris of rock and fallen trees is clearly visible, a scar inching along the eastern shore of the lake. Poppy and Mother can barely look. Their only comfort is that they can neither see nor hear the road from the cabins. It's their consolation prize.

"Your great-grandmother insisted that the road be far away from our cabins," I say.

"Wow!" Fiona says with admiration. "She really fought hard."

"Yeah," says Bronwen. "Imagine if the road were right beside our cabins."

"I'd hate that," says Fiona.

"Me, too."

Two years later, the forestry road is completed. No one in the family is happy. Now we hear logging trucks grinding their gears up the hill at Otter Cove. The trucks stir up dust in the dry summer months. The valleys at the south end of the lake are open to logging. You can see the clearcuts from the south end. Every time I see a clearcut, I shudder. Does it have to be such an eyesore? Reforestation is mandated, but it never happens. I phone and complain to the Forestry Branch, but I don't get much satisfaction. The official gives me platitudes not positive action.

Fiona and I drive into the city the second week in August to do some errands. We take the opportunity to visit Mother at Crofton Manor. I try to bring one of the grandchildren on every visit, for their sake, for Mother's, and for mine. We wheel her into the visitor's lounge. We set ourselves up at a small table. Tea and cupcakes and a game of dominoes. Mother keeps asking me who the pretty girl is. Fiona is patient with her, repeating that she is Michael's daughter, her great-granddaughter. We tell her that we were at the lake last week.

"We caught minnows in front of the Sandpiper," Fiona says. "That was your cabin, right?"

Mother turns to me, "Your house is so big."

That is true. We wanted a place big enough for our two boys and their families. Mother thought the house was too big. We shouldn't have cleared so many trees. She grumbled about the changes, just as Poppy had done when she and Dad built their cabin. I would have thought she'd have learned something from all that, but clearly she hadn't.

Ever since we were married, Patrick and I had had the dream of building at the lake. Patrick loved the place as much as I did. We knew exactly where we wanted our cabin, on the other side of Cupola Lodge, with a view of the far mountains. We went over our plans with Mother and Dad and with Chris, too. We walked the site, showing them the large trees that would have to come down. After much debate, we got everyone to agree. Four years later, our cabin was finished.

With every new building, our family footprint at the lake has grown. Each change provokes the need to balance our own comfort with tradition, ratchets up the tension between old ways and new. Poppy railed against Mother and Dad's modern additions. Mother did the same with ours. At the time, I resented her interference, hated having to explain our choices. I see now that it is hard to accept alterations to something that is close to your heart. Continual change might destroy some essential quality of life. In the old days, we had to rely on each other. We worked together to cut down, chop and stack wood for our stove. Basic sustenance was everyone's responsibility: fishing, picking berries, cooking, cleaning. Our entertainment was a game of cards, singing around the campfire, building forts, listening to a gramophone. Today, we come and go so easily. We can have any new gadget that catches our fancy. We have wifi. We have iPhones, computers and iPads, all of which put us in constant touch with the outside world. We don't talk to each other as much as we used to. Has the essence of our time changed? I don't know. But it might. I only know that now I don't want it to change any more than it has. Have I become the old stick-in-the-mud? I wonder what the future will look like. I have two sons and Chris has four offspring. Will any of them want to build their own cabin? And what about our six grandchildren? And Chris's grandchildren, if and when they arrive on the scene? In another century, the property could be inundated with cabins. So long as they're here, enjoying the lake, I'm okay with this picture. So maybe I'm not such an old stick-in-the-mud after all.

When Gran died, the relationship between Mother and Rowly deteriorated even further. For years, Mother had looked after Gran's finances and managed Poppy's estate. Rowly became more and more unhappy with Mother's way of doing things. They each resented the other's influence over Gran. Rowly felt that he, as the first-born and the only son, was entitled to a greater share of Gran and Poppy's estate. Rowly owned shares in Cupola Estates, which holds the family property. He decided several years before Gran died that he wanted to sell. Mother wanted to get her hands on Rowly's shares. She was worried that he would cause a problem. So she arranged that Chris and I would each buy half of Rowly's shares. As far as I knew, Rowly and Mother were happy with this result. But now I wonder.

Over time, even before Gran's death, Rowly and Mother started arguing over everything. They both said things that could not be retracted. Within months of Gran's death, they stopped talking to each other. Neither would give in and apologize. They were both so stubborn, both determined that they were in the right. I tried to maintain a connection with Rowly. For months, I'd phone him every week.

Finally, he said to me, "Shelley, don't call me any more. I don't want to talk to you."

So I stopped calling. Ten years later Rowly died. Our family didn't find out until three months later—a sad ending for a brother and sister who, although they had their difficulties, were, after all, family. Once upon a time, they were friends. I never called Rowly's partner, Richard, to offer my condolences. I regret this. I should have had the guts to call. Mother's only response to Rowly's death was to bemoan the fact that Gran and Poppy's treasures were lost to her. She would never get Gran's dining room table. I couldn't bear to listen to her harangues.

I hope that I learn from the past. I hope that I can embrace change. I need to accept my relationship with Mother with all its imperfections. My brother Chris and I have talked about family ties. We have vowed that we will never let anything compromise our friendship. We are going to stick together, no matter what. This much we have learned from the past.

THE MACLEODS

Gran and my mother, Frances, were at the lake on their own in the summer of the Big Fire of 1938. Poppy would come in a few weeks. In the meantime, Frances, at nineteen, was capable of looking after things: reconnecting the water pipe if the water pressure dropped in the hose; cleaning and refilling the kerosene fridge; fishing for the next day's supper. Gran relied on Frances to keep the place running. She also could call on Charlie Lindeman, who lived a mile down the beach, to help out in a pinch.

The only communication with the outside world was Charlie's short wave radio. But it only worked sporadically. Once the family was at the lake, they were pretty well out of touch.

That summer, the temperatures in the Chilliwack Valley reached a record high of thirty-seven degrees Celsius. The humidity index registered zero.

On July 17, a fire flares at a ranch along the Chilliwack River fifteen kilometres up the logging railway from Vedder Crossing. No one knows how. With the drought, the whole landscape is a powder keg. Headlines in the *Chilliwack Progress* scream "Fire in Valley out of Control!" Three hundred and fifty firefighters work around the clock to keep it from spreading. A strong east wind fuels the flames. Within ten days, the fire is roaring up the Chilliwack Valley, gobbling up forest as it heads towards Chilliwack Lake.

On July 29, Poppy calls Earl MacLeod, a pilot who lives in Vancouver near the Jericho Beach seaplane station where Earl is a squadron leader with the Royal Canadian Air Force.

"Earl, can you fly me in to Chilliwack Lake? I need to check on the water gauge for the Water Power Branch and I'm worried about my wife and daughter who are on their own at the lake."

Poppy knows Earl MacLeod through his neighbours, the Browns, who live across the street in Vancouver. Earl MacLeod is a relation of the Browns and the three couples often have dinner together. Poppy is frantic for news. Is his family safe? Has his cabin burned to the ground? Has the water gauge been destroyed?

Earl agrees to fly him in. The next day, Poppy drives to Chilliwack and is picked up by Earl at the wharf on the Fraser River near Chilliwack.

Earl is a small man, neat and trim. His khaki shirt is tucked into belted trousers. They climb into the cockpit of the Fairchild seaplane and buckle up. Poppy sits in the co-pilot's seat. They taxi down the river, the single engine revving, the noise deafening. Pontoons slap against the river's chop. The plane picks up speed until it lifts off the water and they are airborne. They fly south to the head of the Chilliwack River valley and turn east to climb fifty kilometres up the valley to Chilliwack Lake.

A pall of brown smoke hangs over the valley, reducing visibility. Through the dusky veil, they see green forests below until they reach Slesse Creek, then devastation. Shocked, the men fall silent at the sight of charred trees and cindered forest floor. The fire in its terrible whimsy has spread its fingers up the hillsides of the valley, blackening one strip of trees while leaving another untouched. The Chilliwack River, a ribbon of green, snakes through the valley of ashes. Stark silhouettes of trees stand like sentinels on the ridges ahead. The sun hides behind a curtain of smoke, as if it can't stand the sight either.

Poppy's stomach is tight with anxiety. He has not heard a word from his family since the fire started. Almost two weeks now. The plane approaches the benchland at the top of the valley. Behind it sits Chilliwack Lake. Earl flies the plane low to the ground across the benchland. The large fir trees on the shore are still standing. Poppy stares down.

A shout from Earl. Poppy catches a glimpse of Charlie's cabin, then farther down the beach, his own cabin. Both are unscathed. He lets out a whoosh of breath. Earl flies the plane to the south end, then banks in a wide half-circle to face north and descends to land on the water. About half way, the pontoons touch down and the plane taxis towards the beach and Poppy's cabin. Figures come into view, running down to the edge of the beach waving their arms. His wife, his daughter, and Charlie—all safe.

Over a cup of tea, Poppy and Earl hear the whole saga. Charlie saw billows of smoke coming up from the valley and walked part way down the trail to investigate. The valley was blanketed in smoke. The sun shone orange-red through the haze. In the distance, he could hear the fire's dull roar.

He thought about trying to start a backfire at the top of the benchland. From his vantage point overlooking the valley, he could see new fires leapfrogging up the valley. He knew that a backfire would be useless. The oncoming blaze already covered the valley bottom, cutting off their only means of escape. Charlie decided their best option was to abandon their cabins and go out onto the water. He hurried back to his cabin, gathered supplies and stowed them in his canoe. He tied the canoe behind his motorboat and motored down the beach to pick up Gran and Frances.

He shouted out as he neared the cabin, "Frances! Fire! We have to go. It's coming up the valley. The trail's cut off. It's nearly at the benchland. We have to take our chances out on the water."

Gran and Frances grabbed food and blankets and climbed into the boat. "How're we for gas?" asked Frances.

"I put in extra," Charlie answered. "And I have some pails for scooping water."

Out they went into the middle of the lake. They covered their heads with blankets soaked with water. They watched as the sun went down red against the smoky sky. The flames burst over the benchlands, whooshing effortlessly up trees a hundred feet high, flaring in the night sky like candles. Flames leapt like acrobats treetop to treetop. Limbs exploded in the air. The roar of the wind reverberated through their bones. Searing heat and smoldering ash bore down on them. The flames leapt across the lake. Animals raced down to the water's edge and started to swim: deer, squirrels, rabbits. Frances was heartbroken to think about all the animals that wouldn't escape the fire. The three huddled together in the boat, encircled by the fire. Now and then they took turns trying to sleep. For a day and a half they bobbed, lying in the bottom of the boat, coughing with the harsh sting of smoke, heads pounding. At the end of the second day, the incessant roaring died down. The fire was burning itself out. Exhausted, covered in ash, lungs aching, they looked out over the lake, the smoke a thick pall around them.

They started the motor and headed to the north shore, steadying themselves against the prospect of seeing burned-out shells where their cabins used to be. When they got close enough to see, they leapt up, shouting with relief. The cabins were standing! The tall fir trees around the cabins were also standing, their bark blackened, but intact. They stepped out of the boat onto ashes that seared their feet.

"Thank heaven you're all right," Poppy exclaimed. "Charlie, how can I ever repay you?"

Earl MacLeod's diary recorded the day's flight:

On 30th of July 1938, I carried out a flight for the federal Water Power Branch that was to be of more importance to myself than I realized at the time. The Water Power Branch had a water gauge installed at the outlet of Chilliwack Lake that was regularly read.

Chilliwack Lake, noted for its inaccessibility, beauty and fishing, had held great attraction to me since my boyhood. My passenger was Mr. Christopher Webb in charge of the federal Water Power Branch in BC. He, because of uncontrolled forest fires in the area, wanted to check the fire, not only because of the water gauge but also because of his privately owned cabin at the lake.

We found that both gauge and cabin had been saved from the flames by the only resident at the lake, Mr. Charlie Lindeman, colourful trapper, prospector and philosopher. With Mr. Webb, I checked the fire again on August 15. I fell even more deeply in love with the out of this world scenery, the fishing and sandy beach of Chilliwack Lake.

Three trips were recorded in Earl MacLeod's flight logbook for 1938: July 30, August 15 and August 19, all of them transporting Christopher Webb to Chilliwack Lake.

The fire of 1938 created the thick forest that I knew as a girl. The fire consumed the dead vegetation on the forest floor. Its heat popped open the cones of the Douglas fir, releasing thousands of seeds to germinate into a second-growth forest. I was born ten years after the fire. Then, the fir trees crowded so close together that I couldn't walk between them. Standing on the verge, I was faced with an impenetrable wall of green. When Chris and I built Fort Red Pine, the trees were as thick as stalks of corn in a field. No one could see us from the cabin. Fifty years later, the forest has opened, the smaller trees have fallen, taken over by moss, and broken down to debris. Now when my grandkids build forts, we can see them from the cabin. But in 1938, it was devastation.

Beguiled by the jewel he had seen, Earl decided he'd like to buy land. Earl and Flora came to Poppy and Gran in October 1938 with a proposal to buy a parcel of the Webb property. Gran and Poppy agreed.

I have always wondered why Poppy agreed to sell to the MacLeods. Surely, it must have gone against his grain. Poppy always talked about the importance of holding onto our land. Over the years, he bought more and more land to provide a buffer around his original parcel. I remember Mother saying she was upset when she learned that Poppy had sold some of their land.

So why did he sell to the MacLeods?

I look through Poppy's files and find the original transfer agreement. It was signed on June 17, 1939, less than a year after the fire, and transfers 2.9 acres from Frances Webb to Flora MacLeod. The payment was four hundred and fifty dollars.

The author's grandfather is standing between pilot Earl MacLeod and trapper Charlie Lindeman in front of the float plane that flew him and his family to Chilliwack Lake, circa 1941. Courtesy of Webb family.

Did Poppy sell for money? He paid five dollars per acre in 1929 and sold for one hundred and fifty dollars per acre ten years later. Although it doesn't seem like much now, it must have been a tidy sum at the time. Poppy was a civil servant with a secure job: somehow, it just doesn't ring true that Poppy was motivated by money.

It occurs to me that John MacLeod, the eldest son of Earl and Flora MacLeod, might be able to shed some light on the question. I phone him at his home in Vancouver and tell him that I am looking for the hidden stories of Chilliwack Lake. I tell him I will be in Vancouver next week visiting my mother and ask whether I can drop in and see him. He invites me for coffee.

John and his wife, Jean, live in Kerrisdale near where my grandparents used to live. John greets me at the door and shows me into their den. His memorabilia is piled on the coffee table. He has several photo albums, files with correspondence, and his father's diary and flight log books. John shows me a diary entry:

In 1939, we acquired acreage from Mr. Chris Webb for a summer place of our own. I arranged for a highly skilled axe man and woodsman, Mr. Ed Marshall, to build our log cabin, which was completed in 1940.

My family will agree that our place there has meant a great deal to us. Charlie Lindeman built a boathouse for us in the spring of 1939 and I decided to get leave in the summer and use the boat-house as a sleeping and camping accommodation for a three-week holiday in July.

"John, do you know why my grandfather sold the land to your father?"

"Not really."

He shows me a copy of his family tree. John was born in 1932, his brother Bob in 1933 and sister Flora in 1936. There are now twenty-one descendants of Earl and Flora MacLeod. John tells me about the building of their cabin while we pour over his photo album. I leave with lots of information but not the answer to my question.

I decide to ask John's sister, Flora. I think she's at the lake this week. I think about Poppy and the MacLeods as I drive back to the cabin. I pull into our access road and step out of the car to open our entry gate. Dad scrounged the red metal bar gate, now rusty, from a logging company down the valley. We keep it closed even when we are at the cabins, with the chain looped around to make it look like it is locked. Stepping out of the car to open the gate, I am enveloped by the smell of fir needles, pungent in the hot August air. Back in the car, I drive down our access road until the road splits: the Mac-Leod's road goes straight and ours veers off to the left. I glance ahead. The MacLeod gate is open. I promise myself to canoe over to see Flora tomorrow.

"Will, do you want to come over to the MacLeods with me?" I ask in the morning.

"Can we look for garbage along the way?" Will replies.

"Sure."

This is his thing. Mr. Keep-Our-Lake-Clean. We get into the canoe. He keeps his eyes peeled, as we paddle.

"I see a can," he says. "I can swim down and get it."

I pull the canoe in a bit so he can jump out. He retrieves the can and tosses it into the canoe.

"There's a plastic bag on the shore, Will."

Off he goes. We carry on, him spying garbage to gather, me encouraging him. By the time we get to the MacLeods, the canoe has three pop cans, four beer cans, and miscellaneous plastic bags and wrappers.

Flora waves from her cabin, "Shelley, come on in."

As Will and I climb the stone path to her cabin, I admire the old log cabin, now weathered a deep, silvery grey. The steep shingle roof extends out over the narrow porch. I show Will where they, too, have outlines of fish penciled onto the logs, with dates and lengths.

"Just like the ones at Chris's cabin," Will says.

We look at one: "Rainbow trout - 17" - July 20, 1942."

"I remember the day Dad caught that fish," says Flora. "The largest he ever caught." She continues, "I have fond memories of your grandparents. I'd row over and they'd welcome me in, pour me a glass of lemonade and listen to my stories. I often think how kind that was of them."

Flora is five when she first rows over in their wooden boat to tell Gran and Poppy that she has learned to row. A few days later, she appears to tell them that she has learned to whistle. If Gran and Poppy are taking their afternoon naps, Flora rows back and forth in front of the cabin until one of them appears on the porch. Then she rows in and proudly shows off her latest skill.

Flora likes to sit on the steps of the porch while Poppy shaves. He sharpens his straight razor with a leather strap, lathers up the soap in his bowl and smoothes it on his face one section at a time. Peering into the small mirror hung on a hook on the porch post, he carefully scrapes his whiskers away. When he's finished, he towels his face clean and turns to her with a flourish. Flora applauds. Gran brings Flora and Poppy glasses of lemonade.

"Oh, I nearly forgot," Flora says. "I found a diary of my mother's early years here. I thought you might be interested."

She sits beside me and together we open the worn leather notebook. The writing is cramped, as if her mother was worried she would run out of room on the page. I read out to Will the entry for 1939.

> Lake with the Webbs for several weeks holiday. After enjoying the grand hospitality of Chris and Frances, we set up camp in the boat house at Delhaven Lodge. Earl built an oil house. John caught his first fish. Allan Hull came in unexpectedly a few days before we were due to leave to say Earl was recalled—War. We broke camp in 2 hours. Frances helping me so much and all returned in the Stranraer.

"This really brings your mother alive, Flora," I say. "I never knew she called your cabin Delhaven Lodge."

"The name didn't stick."

"What is a Stranraer?" asks Will.

"It's a bi-wing two engine plane with pontoons to land on water," says Flora.

"Let's look it up, Will, when we get home."

Will and I say goodbye to Flora. As we leave, Flora says to me, "You remind me of your grandmother, Shelley."

Earl and Flora MacLeod in front of their newly completed boathouse with their son John, circa 1939. The MacLeod family acquired a parcel of land on Chilliwack Lake from the Webb family in 1939. Courtesy of MacLeod family.

Flora can't possibly know the joy I feel in hearing those words.

Back at the cabin, Will and I look up the name. A Stranraer was a flying boat built in the 1930s for the Royal Canadian Air Force. We look at a photo: it has two levels of wings joined by metal struts, with two propellers at the front. The hull is covered with sheet metal and the wings with fabric.

"Fabric? Like my shirt?" he says.

"Yeah, it says here, the earliest planes used cotton," I say. "Kind of like your shirt."

"Are you sure? Doesn't sound strong enough."

Throughout the afternoon, I ponder Flora senior's diary entry. It shows a friendship between Flora senior and my Gran. I mull over the timeline of events. In 1938, Mother was nineteen. She was an undergraduate at the University of British Columbia but living at home. German aggression in Europe was escalating. War in Europe was imminent. But I doubt these world events, although crucial at the time, would have influenced Gran and Poppy's decision to sell off a piece of their land.

I think it was the fire.

I listen again to Kevin's tapes and hear Mother's voice:

The fire leaped from tree to tree. We didn't think we would make it. The fire came right to the edge of the cabin.

Gran and Frances were terrified. The fire would have made Gran feel vulnerable. She was not completely comfortable at the lake in the first place. She never stayed there on her own. Neighbours would offer security. Help close at hand. The companionship of another woman. I check the MacLeod family tree: Flora MacLeod was thirty-seven in 1938. Gran was forty-five. Close enough for a friendship to form.

Maybe this was Poppy's way of saying thank you. Earl had flown him to the lake to see that his family was safe. It speaks to Poppy's sense of honour, a debt was repaid. And as a bonus, Earl offered the potential of transportation.

Earl MacLeod ushered in the era of flying to our mountain retreat. After the fire of 1938 destroyed the logging railway, Gran and Poppy had had to travel the last thirty kilometres by horseback, a ten-hour journey. By contrast, the trip by float plane from Chilliwack took less than an hour. Once the Mac-Leods built their cabin, the Webbs joined the MacLeods on their flights to the lake for their summer holidays. Gran and Poppy never again came in by horseback.

For the first time, we had another family. Another couple with young children. Gran and Poppy could row over to visit, have a cup of tea, and stay for lunch.

At the beginning, the two families' lives were quite similar. We both had simple log cabins and came for a single three-week stretch every summer. Over time, however, our approaches diverged. The Webb family footprint grew: our family built two more cabins and added onto the original cabin so that all of us could be together at the lake at the same time. The MacLeods did not have the room to expand. Instead, they shared the same cabin Earl and Flora built at the beginning of the war.

We were a ten-minute row from the MacLeods, but even so, we were attuned to each other's comings and goings. As a girl, if my ears picked up the sound of an approaching plane, I was immediately on alert. Was the plane coming for us or for the MacLeods? If the plane headed for the MacLeod beach, I'd hop into the rowboat and head over to hear the news. There was a gap in the children's ages between the two families. By the time Chris and I were born, the only MacLeod child still at the lake was Flora, who was twelve years older than me. The age difference mattered at first but over time Flora and I became friends.

One of the common experiences that we shared with the MacLeods was the annual trip to Kokanee Creek for the spawning run. Although we didn't go together, both of us remember the creek red with fish from bank to bank and its icy chill. By the time Chris and I were old enough to come along, John was married in Victoria with two young children and Flora was away.

It is August 4, 1958. We set off from our cabin at daybreak in a variety of motorized craft, canoes and fishing boats. Ray Wells leads the flotilla in his aluminum boat with a dent in the prow where it was dropped by the helicopter that flew it in some years ago. Poppy, Mother, Dad, and I are in our wooden fishing boat with a ten-horse motor. Chris, who is eight, has hopped into Ray's boat, in the hopes of hearing some of Ray's stories during the hour-long trip. Gran has stayed behind for a quiet day. Once we get to Kokanee Creek, we pull up the boats on the sand at the creek mouth and tie them to trees. Chris and I have on T-shirts, shorts and sneakers and shiver in anticipation of the bone-numbing cold of the creek. We take out our short-handle nets and canvas fishing bags.

"Off you go," Poppy says. "Let's see who gets the most kokanee."

Poppy has long schooled us in the ways of the kokanee, a fresh-water salmon that returns to its spawning ground in its fourth year. When spawning, the kokanee turns crimson red. The female kokanee turns on her side, making an indentation in the gravel by flexing her tail. With the male at her side, the eggs and sperm are shed together into the indentation, then the female covers the eggs over with gravel. After laying several batches of eggs, the female dies.

I am ten. I step into the creek, wincing with the cold. I keep my foot in as long as I can, then bring it out of the water, tingles shooting up my shin. I try again. Each time I can keep my foot in longer until I no longer feel the stream's icy embrace. Feet numb, I splash up the creek with abandon. The kokanee spawns in the gravel shallows at the mouth of the creek and travels upstream to die. The creek is narrow with thickets of elderberry choking the edges that scratch my arms and legs. The water of the creek is a mottled icy green and brown. I search for a glint of red, and suddenly I see a kokanee in the lee of a tree fallen across the creek. Its forked tail creates ripples with an occasional crest that foams above the surface. I follow, trying to anticipate its next move. Will it dart into the deeper reaches or dash forward? I plunge my net into the water and come up with my first kokanee. I whoop and catch my brother's eye as he looks up. A bang on its head with my stick and I lay the kokanee in my bag. Its skin shimmers, crimson with speckles of black and silver. I rinse my slimy young hands in the creek and continue upstream.

Another flash of crimson among the brown, grey and white stones on the creek bottom. The flow of the creek is fast, water rushing over rocks and fallen trees, eddying with ripples and curls of frothy white. I reach in again with my net. I miss.

Hours later, we gather at an open spot alongside the creek. Dad lights a fire and we fry kokanee in skillets for lunch. Late in the afternoon, we gather back at the beach and pile our nets and bags into the boats. The kokanee are laid in coolers. The return trip against the wind takes nearly two hours. Our ten-horse motor labours against the waves for the six-kilometre trip, the bow of the boat lurching forward, then cracking down hard on the surface of the water. We hunker down under old overcoats, shivering from the cold spray of the waves.

When I think back on our annual trip to Kokanee Creek, I wonder what on earth we were doing. Did we net fish before or after they had laid their eggs? Did we interfere with the spawning? I have no idea. This practice borders on barbaric. It certainly was thoughtless. At the time, no one considered our impact on the spawning kokanee. I remember that Ray Wells was insistent that we not waste any fish. Ray had a smokehouse and smoked most of the harvest for the winter. Our family just kept enough for dinner the next day. It was sport. It never occurred to us to question what we thought of as a fun day's outing. Today, we would be pondering such questions. Fifty-six years ago, we did not.

Throughout the summer months, Earl and Flora senior travelled back and forth between their cabin and ours by rowboat to visit my grandparents. Although there was a ten-year difference in the ages of the two men, they were similar. Both were neat and tidy, precise in manner. Earl had an arresting voice and spoke with authority, someone who was used to deference, eventually rising to air commodore in the Royal Canadian Air Force. Both women took care with their appearance for these visits, donning a felt hat with a feather in the band or pinning a brooch onto the collar of a shirt. As if they didn't want to let standards slip at summer camp. Flora sat erect in the front of their rowboat, like a queen being ferried to greet her subjects. Mostly, they came for lunch, then headed home for their afternoon naps.

Despite the deaths of the patriarchs—Poppy in 1974 and Earl in 1987—the following generations of the two families continued the close bond forged by the grandparents.

We carry out projects in common. Flora and I confer about logging and mining plans for the Chilliwack Valley. Flora writes letters raging against logging practices that scar the hillsides.

A few years ago, we went to the MacLeods with a proposal to bring power to our cabins. We had been using solar backed up by a diesel generator for twenty years. The MacLeods had been using propane for lights, fridges and stoves. Our solar system could not keep up with the demand once Kevin and Michael had families. Balancing these systems to keep up with increased usage was difficult. And we wanted to make the operation of the cabin simpler for the next generations.

Once BC Hydro put power into the provincial park, we started to think about bringing power across to our land. We approached BC Hydro and came up with a plan to put in power poles down the forestry road from the provincial park to our access road, then bury cable down our access road to our cabins. The MacLeods agreed, and we completed the project together. For the past four years, we have had electricity. Chris stayed off the grid and continued to rely on his solar system. He lives in Gran and Poppy's old Cupola Lodge, so it's not surprising that he wants to keep his life old-style, with as little change as possible. We both understand that we have a difference in approach to life at the lake. I want convenience; he wants tradition. We've learned to accommodate each other's views.

My mother was against the sale to the MacLeods. She wanted to keep our land intact.

"If I'd been around, the sale never would have happened," she railed to me later. I think she cast herself in a misleading role. She was only nineteen at the time. She would have had no say in the matter. I'm glad Poppy sold to the MacLeods. It seems right, given the dangers Earl faced to unite Poppy with his wife and daughter during that fateful fire. They have been good neighbours. Now someone else has the same history and ties to the lake. They share our devotion. The lake is their favourite place on earth, too.

Church Camp

It is a hot and sunny day in mid-August.

"Let's have a picnic down the lake," I say at breakfast.

"Yes!" a chorus from the rest of the family.

Our family loves picnics. We always have. What is it about a picnic that draws us in? I like the time alone with the family, especially the grandkids. We have no distractions, no chores to do, no diversions. Lying on a beach towel in the sun invites all sorts of conversations. Just staring up at the clouds sparks the imagination.

"Hey, there's a cloud shaped like a boat."

"I see an elephant."

The preparations happen beforehand. All we have to do is open the hamper and eat. And the kids love picnics. I only have to say the word and they jump up with enthusiasm every time. And we always bring back a story.

"Dennis knocked over the lemonade, spilling it all over the sand."

"Murphy rolled in a dead fish, then sat on my towel."

"Ants crawled on the sandwiches, so we couldn't eat a thing."

"The canoe tipped over. We got sopping wet."

Picnics are the stuff traditions are made of.

This morning, we pack the picnic basket, make up the lemonade. The kids are already in their bathing suits. They gather up their towels hanging on the porch railing.

Michael asks Roger and Fin to bring the cans from the gashouse to the end of the dock so that he can fill the tank of the motorboat. Once we get the "all clear" wave from him, the rest of us straggle down the stairs, across the beach and onto the dock. One by one, we step into the boat and settle in our places. I choose my usual seat on the captain's side of the bow, in the sun and close enough to chat to Patrick. Murphy sits like a canine figurehead on the cushioned seat at the prow. Patrick is at the helm.

"Where are we going?" asks Nora.

"Our usual tour," says Patrick. "We'll stop at Paleface Creek beach for lunch."

"That's the one you used to call Kokanee Creek, right, Rani?" says Fiona.

"Good memory," I say, pleased that the kids are listening to me when I tell them the stories of the lake. Turning to Patrick, "Let's stop at the Church Camp on the way. I've got some questions for Kris Sanders about the camp's history."

"Okay," says Patrick.

"Can we go to the Jumping Rock?" asks Fin.

"Sure."

Patrick puts the boat into gear and we slowly motor along the north shore of the water. The breeze is imbued with fir and cedar. Heat reverberates off the surface of the water, distorting the air. When we arrive at the Jumping Rock, Patrick brings the boat in gently. Michael steadies the bow against the rock face as Kevin points out the best route for climbing up. I get out the paddle, ready to fend the boat off the granite surface. One by one, each of the grandkids clambers up the rock with a helping hand from the dads. Joanna pushes us off once they're up. Patrick, Joanna, Molly and I motor out a distance from the Jumping Rock and turn off the engine to watch. Molly has her camera ready. The kids and dads stand in a line, talking about who will jump first. They arrange themselves in order of height, with the tallest at the highest end, about three metres off the water.

"Let's all jump together."

"Okay, let's hold hands."

Eventually, someone shouts, "On three. One! Two! Three!"

They jump, arms waving like windmills, bathing suits flapping. Eight splashes in a row. Heads pop up, wet and sleek.

"The water went up my nose!"

"Ack! My bikini bottoms are down at my ankles."

"Let's go again!"

They swim back to the rock for another jump. When they're ready to get back in the boat, Patrick cuts the motor and Molly puts down the ladder at the stern. As each one climbs aboard, Joanna passes them a towel.

"How was it?" I ask.

"Awesome!" says Roger.

We carry on to the beach that stretches across the entire south end, the inrush of the Little Chilliwack flowing from the Cascade Mountains in the United States creating islets at its mouth.

"Remember when we found the silver mine at the beginning of the summer?" says Bronwen.

"And we walked down the creek. That was crazy," says Nora.

"I want to hike in there again," I say. "The miners apparently had a cabin. Maybe we could find remains along the river bank."

"Yes!" says Fin. "Another adventure!"

The water over the shallows is pale green. We look down from the boat and see schools of Dolly Varden, an inland char with light speckles on its olive-green back. I wonder how the fish got its name? I'll try to remember to look it up when we get back.

We turn the boat and continue up the east side. We pass where we brought the boat in to search for Michael Brown's grave.

"Let's go see the grave again," says Roger.

"No, not this time. But I'd like to come back to see if we can find any pieces of the wooden cross," I say. "Maybe later in the fall?"

"I want to come," says Will.

"Yeah, me too," says Fin.

Further up the east side, we come to the Church Camp in the bay south of Paleface Creek.

"How long has the camp been here, Rani?" Nora asks.

"For as long as I can remember." But that's not enough of an answer. The Church Camp has always been a fixture at the lake, but when did it open? Who started it? Why put a camp at such a remote location?

So many questions still to answer and time is speeding by. Only two more weeks until the grandchildren head back to school. I am surprised by how much I am enjoying the research. It is like working on a jigsaw puzzle: I hunt and hunt for the right piece of information and, when I find it, I fit it into what I already know, gradually developing a complete picture.

Last week, I started with an internet search for the Church Camp. Almost immediately, I came across the Chilliwack Lake Foursquare Camp on Crossroads Youth website.

> A rustic volunteer run log cabin lodge located at the east end of Chilliwack Lake Provincial Park. This beautiful facility has the ability to accommodate more than 50 people in two buildings. Both buildings are generator driven, and there are no showers on site. Although this is not a 5 star hotel, this location enables us an opportunity to enjoy the sandy beach, and glacier fed lake, giving us an incredible outdoor experience similar to camping but in a much warmer and secure environment!"

I discovered another entry on the Hearts of Fire website, *All about Hearts of Fire Summer Camp.*

> Hearts of Fire Summer Camp is a camp run by Christians for youth who may not be able to afford camp. We run 3 consecutive camps in

August each year. The camp, which is owned by Four Square Church, is held at a Christian camp at the far side of Chilliwack Lake.

I don't know anything about Four Square Church. Does it own the Church Camp? I find a website for Canadian Foursquare Church. I submit an information request. A message left on my phone suggests that I contact Kris Sanders for information on the Church Camp. I've come full circle: the answer is here at the lake.

I have already talked to Kris about the location of the silver mine and Michael Brown's grave. Clearly, I need to go back to him to find out what he knows about the history of the Church Camp. I give Kris a call. It turns out he is the son of one of the volunteers that worked on the camp in its early days.

It all started with Reverend George Wright who founded a Pentecostal church, Zion Tabernacle, in East Vancouver at the end of the Second World War. He opened an orphanage and started a club for boys whose fathers were killed in the war. Kris's father, Gerald Sanders, was one of several men who assisted Reverend Wright with the Boys Club and the launch of the Church Camp. Reverend Wright had spent time at Chilliwack Lake in 1916 as one of Chief Sepass's partners in the Silver Chief Mine. Its beauty and solitude stayed with him: it would be the perfect place for a boys' camp.

In the summer of 1946, Reverend Wright and Gerald Sanders hiked in to the lake with a group of boys. Packhorses carried provisions and forded the creeks. One horse carried a boat on travois poles held up in turns by the two biggest boys.

The first summer, the boys built a boathouse at the north end beside the MacLeod's, where they could lock up their boat for the winter. They stored food and supplies in Charlie Lindeman's abandoned log cabin at Paleface Creek. They slept in tents. Through the summer, they worked cutting and peeling fir trees, and by the end of August, they had built a lodge.

That first visit took place two years before I was born. By the time I was six, the comings and goings of the Church Camp were part of my summer routine. By the time I was ten, I was intrigued.

It is early August 1958 and I am on the lookout for signs of activity on the beach beside the MacLeod's cabin. The horse trail from the valley ends on the bluff between the MacLeod cabin and Charlie Lindeman's. There won't be a road into the lake for another three years. I know that the boys from the Church Camp will be here any day now, riding in on horseback. I can't wait.

Transporting a wooden rowboat by a travois to the Church Camp, circa 1939. The boat arrived in one piece at the north end of the lake and was put on a raft and motored down the lake to the Church Camp. Courtesy of Kris Sanders.

Why am I interested? Am I bored? I only have Chris, two years younger than me, as my companion. These boys are older, maybe twelve or thirteen. Definitely interesting to a ten-year-old.

One afternoon, I hear shouts and the banging of tin boats. I row over with Chris, wave hello to the MacLeods and sit on a log watching the action. Above, on the bluff, ropes are loosened, boxes and sacks and duffle bags lifted off the horses' backs. Boys carry supplies down the slope to the beach. Earl MacLeod wanders over to talk with Reverend George Wright and Gerald Sanders.

The doors of the Church Camp boathouse, padlocked during the rest of the year, are wide open. It's a simple structure, with boards walling in the sides, and a pitched roof overhead. From the open doors, planks extend towards the beach. Each of the camp's boats rests on wooden rollers. One by one, they are rolled down the planks and onto the sand. A couple of the boys, small and nimble, are transferring the rollers from the back of a boat to the front so that each boat can continue its movement forward down the beach to the water. At our cabin, we use this same roller method for our boats.

I check out the boys. They are rough and tumble. They ignore me. Would I like any of these boys if I had a chance to get to know them? Mother says the boys come from a poor background. I don't really know what this means. I know that I go to a good school and live in a nice house. My parents don't talk about money but I think we have as much as we need. These boys probably don't. I am awkward, feeling I am not the same as them. I'm shy. Mother told me not to fraternize, whatever that means. In a way, I'd like to talk to them. But I wouldn't know what to say. It's easier to watch from the sidelines. Chris, on the other hand, is Mr. Sociable. He's gregarious, and friendly. He has no trouble going up to the boys to say hi.

Soon the boats and rafts are packed and the boys divvy themselves up among the boats. They push off down the lake, heading to the Church Camp's site at Paleface Creek. It's a rag-tag bunch, with three or four boats towing a couple of rafts piled high with boxes and bundles. Their voices singing camp songs fade into the distance. A part of me wishes I were going with them. I wish I knew more about their life at the Church Camp. Is it the same as mine at our cabin?

I still want to know.

The Church Camp boys hiked in and came on horseback until the road to the north end was completed in 1961. After that, they travelled by car. But they still had to cross the water like a band of floating gypsies to get to the Church Camp. In 1969, a single-lane logging road was completed down the east side of the lake to Paleface Creek. For the first time, they could drive right to the camp door.

I am at the cabin on an early July weekend in 1969, taking a break from my summer job as a teller. I haven't paid much attention to the boys' comings and goings for a while. I am twenty-one, more interested in college boys than campers. The sound of a chainsaw breaks the morning silence. Curious, I head out in the canoe towards the MacLeod's to see what's going on. As I come closer to the commotion, I see several men milling about the Church Camp boathouse and boys dragging newly cut poles out of the forest. The water is high and laps at the front of the boathouse. I paddle closer and ask what's happening.

"The road's finished. We can drive to camp now," one boy responds.

"We don't need the boathouse here any more. We're taking it to camp."

"How're you going to do that?" I ask.

"We're going to float it down the lake."

I stay and watch. I want to see what will happen next. I wonder what this change means to me. It's part of the summer routine to have the Church Camp boys set off from our end by boat to their camp at the other end. Will I miss that? Or will we welcome the quiet?

Along with the boys are five men, the camp leaders, I think. The men hoist each side of the boathouse and slide logs under the structure. With poles, they lift the back end of the boathouse and jimmy it forward along the logs. Slowly, the boathouse slides forward into the water until it's afloat. They tie ropes from each side of the boathouse to two motorboats, careful that the ropes are exactly the same length on each side. Then they climb into the boats and motor slowly down the lake, the boathouse trailing behind them.

The following weekend I am back. Curious whether the boathouse made it in one piece to its new site, Chris and I take the motorboat to visit the Church Camp. As we round the corner of Paleface Creek, we see the boathouse set up on the shore, none the worse for its crossing.

According to Kris Sanders, Reverend Wright led the Church Camp operations until his death in 1965. His son Wendell inherited the camp and continued his father's work. As a young man, Wendell had attended Foursquare Bible College in Vancouver. In 1984, just before his death, Wendell gave the Church Camp to Foursquare Gospel Church of Canada. It continues to own the camp but each group that uses the Church Camp administers its own programs.

Gerald Sanders was dedicated to the Church Camp throughout his lifetime. His son Kris, who was a camper as a boy, leads the group of volunteers who run the camp now. Most of the volunteers are descendants of the boys who were camp boys in the 1940s and 1950s. Over time, however, the focus has changed. No longer a camp just for fatherless boys, it can be booked by any non-profit community group, now mostly youth groups.

I wonder about the Church Camp's future. The volunteers have

Pushing the boathouse off the beach to ferry it down the lake to the Church Camp, circa 1969. Once the road was completed to the south end of the lake, the Church Camp could be accessed by car. Courtesy of Kris Sanders.

Boys returning from a three-day hike to Paleface Mountain, slogging through Depot Creek on the way back, circa 1958. The Church Camp was a single room log cabin with a surplus army tent for sleeping. Courtesy of Kris Sanders.

a historical connection to the place and are committed to keeping it going. What happens as these volunteers age and can no longer help out? In twenty years, when Kris is in his eighties, who will volunteer to run the camp?

I didn't know Kris when he was young, yet he spent a lot of time at the lake, just as I did. He's maybe ten years younger than I am, so we would have been there at the same time. But, from the start, we didn't have much communication with the campers. Other than a brief visit when they arrived at our end, we didn't mingle. Once the campers were able to drive in by road, we didn't even have that brief connection. And after I was married and had children, although we came to the lake each summer, we never dropped in on the Church Camp. Perhaps as our time was a brief few weeks, we wanted to savour every moment on our own.

Kris initiated the contact. One day about six or seven years ago, he pulled his motorboat alongside our dock, tied up and came up to introduce himself. I'm glad he did. Finally, I was able to put a face on the Church Camp. He was interested in getting a phone system hooked up. He wanted communication in case of emergencies. By that time, we had installed a satellite system. He had worked for Telus and was investigating options for a landline. We chatted about the pros and cons. He gave me contact information for the Ministry of Forests and the Parks Department. He and I started emailing each other whenever one of us heard any news. Now, we drop in on each other whenever we're at the lake.

Not only is his history similar to mine, we share the same concerns. Before a few years ago, I had thought of myself as the only crusader of my generation. I was wrong. There's Kris. For years, he's been keeping in touch with the forestry officials, providing comments on logging plans, championing reforestation.

Kris is a font of knowledge of lake history. Clearly, he's been gathering information for a lot longer than I have.

"We were squatters, you know," he says.

"Same with us," I say. "We didn't buy our land until six years after we built the cabin."

"Well, you bought your land. We only had a lease."

The camp obtained a park permit from the Ministry of Forests to use Crown land for the camp. A bunkhouse, outhouse, wharf, woodshed and gashouse followed the lodge in quick succession. A line of logs was chained together beside the wharf to divide off a protected area for swimming. Because it never owned the land, the camp relied on authorization from various provincial departments to operate on Crown land. In 1973, the Forestry Branch opened a campground at the mouth of Paleface Creek right next to the camp. Curious provincial campers wandered across Church Camp property and peeked into cabin windows. Kids brought their towels and swam off the Church Camp dock. Pop cans and plastic bags floated into their beach. This intrusion lasted for twenty years. In 1994, when BC Parks closed the campground, the Church Camp breathed a sigh of relief.

In 1997, the Chilliwack Lake Provincial Park was extended to encompass all the mountains encircling Chilliwack Lake. The new boundaries included the Church Camp. Kris was advised that the camp would not be able to stay within the new provincial park. He rallied his volunteers and they lobbied the Ministry of Parks. They wrote letters and visited parks officials. In the end, the Church Camp was allowed to stay: it was given a ten-year park lease for 5.67 hectares of land next to Paleface Creek.

Not all provincial initiatives were met with resistance. Kris was enthusiastic about the establishment of the ecological reserve by BC Parks in 1980. On the BC Parks website, I click on the background document on the Chilliwack Lake Ecological Reserve. The reserve is located on the floodplain of the Little Chilliwack River. An area of eighty-six hectares of productive floodplain forest was set aside for the purpose of research and fish stock improvement. It is described as a site of exceptional tree diversity and growth, and of hybridization between coastal and interior conifers. Stands of old-growth redcedars of exceptional growth reach three metres in diameter. I remember being awed by these huge redcedars on our trek to find the boundary markers earlier in the summer. One of the threats listed in the background document catches my eye:

> The reserve is located on the US-Canada border. Drug smuggling and illegal alien crossing is suspected to occur via the reserve.

I have often wondered whether nefarious activities occurred across the border that is practically within our line of sight. Surely, the border is too remote to be practical for running drugs. But what do I know about being a smuggler? Idly, I search for newspaper articles on drug busts in the immediate area. I go to the *Chilliwack Progress*. Nothing. But an internet search shows up the following on the *Seattle Post-Intelligencer* website in 2004 under the heading, "In the Northwest: Smugglers hiking new routes in parks, wilderness."

> The lookout at remote Copper Mountain—one of North America's great mountain viewing points—spotted "unusual aircraft activity" far below in the Chilliwack River valley. A road extends down Chilliwack Lake in Canada to within about a mile of the U.S. border. It was built so B.C. loggers could clearcut the Depot Creek valley to the border of the U.S. national park. Smugglers were hauling B.C. Bud (marijuana) across the 49th Parallel and up the Chilliwack River to a rendezvous location. Bags were dumped for pickup by a helicopter on the American side.

Oh my god! Is this for real? I shudder at the thought of smugglers on the lake, even at the far end. Was this a one-time thing? Or does it happen all the time? I have never even contemplated that smugglers might be at the lake. Every now and again, a visitor, when hearing the United States border is only three kilometres away, remarks, "Wow! Would be a great place for a smuggling ring."

We laugh as if that's the most absurd thing we've ever heard. Now I know it's possible. I hate the idea. It contradicts everything that our retreat stands for: peace, tranquility, harmony. I don't want the outside world intruding on my paradise, especially not drug dealers.

I remember with a jolt—the dead body. About ten years ago, my brother Chris and my nephew Blake found a body on the east shore of the lake, a few kilometres from our beach. It was December and they had come to check on the cabin. They were paddling in their canoe and came around a bend when Chris saw a form on a large rock on the shore. As they got closer, the shape became clearer.

"Oh my God, Dad," said Blake. "Is that a body?"

They went in for a closer look. There was no movement. They stepped out of the dugout and came around the rock. It was a man, about forty years old. He was wearing pants and a blue denim shirt. No jacket, no shoes or socks. He was face up, his arms splayed outwards.

"He looks so peaceful," said Chris. "Like an angel has placed him gently down on the rock."

"Look, Dad. There's a whole bunch of American hundred dollar bills."
Bills were crumpled on top of his body. His face was discoloured but there
was no bullet wound or sign of struggle.

"Don't touch anything," Chris said.

"What'll we do?"

"Leave everything," said Chris. "We'll drive into Post Creek and phone
the RCMP."

They paddled back, questions swirling in their heads. Who was this
guy? Was he murdered? Was it a suicide? Or an accident?

The Royal Canadian Mounted Police officers asked Chris to meet them
on the forestry road so that Chris could show them where the body was.
As Chris and Blake were waiting, a small decrepit car drove slowly by. The
swarthy man driving gave them a hard look. Chris was rattled but managed
to take down his licence plate number. It was the middle of winter, no one
around, and a creepy guy in a rust bucket just happened to drive by?

When the police arrived, they drove together down the forestry road
three or four kilometres along the side of the lake. Falling snow caught in
the glow of their headlights. Suddenly, they saw a pair of bedroom slippers
in the middle of the road. The feet were facing them and the backs of the
slippers were folded over, as if their owner had slipped out of them quickly.
They stopped the car and got out, looking over the side onto the lakeshore
below. The body lay there.

"He was killed, right?" asked Chris.

"Oh, yeah, he was killed alright," the RCMP officer replied. "You can go
home now, we'll take it from here."

Chris didn't hear a thing until a few years ago he got a phone call from
the RCMP officer.

"You remember the body you found at Chilliwack Lake?" he said. "Well,
the victim's sister wants to talk to you. She's having trouble accepting her
brother's death and wants to talk to the person who found him."

Chris called her. He learned that her brother was East Indian, and thirty-eight years old. The RCMP never found out who killed him.

"What did he look like?" she asked.

"He looked peaceful," Chris replied.

I wasn't at the lake. Chris told me about it later. Like the drug smuggling,
I push it to the back of my mind: my usual response to things I don't want to
deal with. Some things I'd rather not know.

As we motor towards the Church Camp in the boat, we see a man waving us in from the end of the wharf. It's Kris.

"Come on in," he calls, wiping grease from his hands onto his coveralls.

We pull up to the wharf, and everyone climbs out. Kevin ties up our boat and we stand on the wharf chatting.

"Can we go and explore?" says Roger.

"Sure, go right ahead," says Kris. "There's only a few of us here now, doing repairs and getting ready for the next camp group."

The kids wander off in the direction of the playground behind the main lodge.

"I'm trying to fix our generator," Kris says. "We were broken into last week and someone vandalized the utility shed."

"That's a drag," says Michael. "I hate it when we get broken into."

Break-ins come with the territory of owning a remote cabin. We've had several over the years, often in the winter. The road is not ploughed and there is anywhere from a skiff of snow in December to two metres by February. Patrick and I have snowshoed in and stayed a few days, the only sound the creaking of branches and lapping water. The lake doesn't freeze over; the shoreline ice extends no more than three metres into the water. The sun rises late and sets early between the mountain ranges, with only a four-hour window of direct sunlight. The colours are shades of white and black. Except when the sun shines. Then everything sparkles and the blue of the sky hurts your eyes.

In 1968, someone broke in and lived in Cupola Lodge for the winter. They used the bedding and pots and pans. We only found out about these squatters when we arrived in the spring. By then, they were gone, leaving a mess in their wake.

The worst break-in at Cupola Lodge was in early September 1974. Someone smashed through both gates on our access road and broke down the door of Poppy's cabin. All the smaller furniture was taken; tables, chairs, lamps, a burl stool that Charlie Lindeman had made, and the gramophone. I was living in Toronto at the time, about to give birth to Michael. I hated the thought of losing those treasures, none more so than the gramophone that I listened to so raptly as a girl. Dad thought the thieves must have brought in a truck. He called the RCMP in Chilliwack and they came out to take a look. They were not optimistic.

"It looks like someone was outfitting their cabin," the officer said. "I doubt we'll find the thieves."

My parents decided not to tell Poppy. Poppy was eighty-seven and was laid up with bronchitis. They didn't want to upset him. He was admitted to the hospital and died a few weeks later. I am glad he never knew.

In the spring of 1992, I got a call from the RCMP. A passerby had reported that a cabin at Chilliwack Lake had been broken into. Kevin was away at university, Michael at a school field hockey trip, Patrick away on business. My brother Chris and I drove up together from Vancouver. Walking into my cabin, Creekside, my heart sank. Drawers were upended, contents strewn. This was no bear: the fire extinguisher had been sprayed throughout the cabin. Knives thrown into the log walls. My dismay turned into anger. How could anyone do this?

Scrawled across the front page in our guest book, "Grad Weekend, Brookswood High School." The police spent a week interviewing students at the Langley secondary school but learned nothing. No one was talking. Michael and his friends were incensed. They wanted to find these thugs and beat them up. They cooled down when they realized there really was nothing they could do. For the next few years, the RCMP set up roadblocks on the Chilliwack Lake Road over grad weekend and confiscated liquor they found in teenagers' cars. The police presence made a difference. We've had a few minor break-ins since but nothing on this scale.

These memories flash through my mind as we're talking to Kris on the wharf. Kris tells us about the Church Camp's worst break-in. In December 1988, a fire set by vandals burned the lodge to the ground. There was nothing left but metal door hinges and the wood stove.

"Oh, I can't imagine our cabin being burned to the ground," I say.

"We'd be devastated," says Patrick.

Kris says he managed to drive in through the snow and secure the sheds and tidy up. He and his volunteers raised money over the winter and rebuilt the lodge the following summer.

We say goodbye to Kris after a half hour of exchanging news and stories of the past. After leaving the Church Camp, we pull up the boat onto the sandy spit at Paleface Creek and lay out our picnic. From here, Mount Lindeman looms large across the water. A craggy castle tower perched above a triangular glacier. The kids go off to explore.

"That would be awful, to have the cabin burned down," says Joanna.

"Yeah, I don't even want to think about it," says Kevin.

"Hey, come over here," shouts Roger. "Fin and I found some marbles in the sand."

The kids run off down the beach. They gather about fifty marbles along the water's edge. What are marbles doing here? A treasure to them, a sign of intrusion to me. The older I get, the more possessive and protective I am. I have the duty to pass down the lake in the same pristine condition in which I experienced it as a girl. Finding marbles left by some casual passer-by irks me. The marbles aren't toys; to me, they're a sign of disregard for the environment. Like finding graffiti scrawled on the rocks.

"CM loves BH" inside the shape of a heart.

Graffiti makes me angry. Patrick bought dark grey spray paint that we spray over such defacements. I must remember to show Will where the spray can is so he can take over this chore. I think he'd like that.

On the Paleface Creek beach, we sit on blankets and pass around the picnic staples: ham sandwiches, lemonade, carrot sticks, cut apples and Oreo cookies. A few wasps join us. The kids jump up, not wanting to be stung. The adults swat them away with their T-shirts. It's afternoon; the wind is building on the lake. We finish up our picnic and stow our things back onto the boat. We head for home. The boat bangs against the oncoming waves. The kids are wrapped in towels, heads nodding in the sun as we make our way back to the cabin.

"Can we have a campfire tonight, Dad?" Bronwen asks as we arrive back at our dock.

"Great idea," Kevin says. "You kids can help gather some wood."

Back in the cabin, after I put the picnic things away, I decide to look up Dolly Varden on the internet. I read the excerpt on Wikipedia aloud to the kids who have gathered around, attracted by the glow of the screen.

> The first recorded use of the name "Dolly Varden" was applied to members of *S. confluentus* caught in the McCloud River in northern California in the early 1870s. In his book, *Inland Fishes of California*, Peter Moyle recounts a letter sent to him on March 24, 1974, from Valerie Masson Gomez:

> My grandmother's family operated a summer resort at Upper Soda Springs on the Sacramento River just north of the present town of Dunsmuir, California. She lived there all her life and related to us in her later years her story about the naming of the Dolly Varden trout.

> She said that some fishermen were standing on the lawn at Upper Soda Springs looking at a catch of the large trout from the McCloud River that were called "calico trout" because of their spotted, colorful markings. They were saying that the trout should have a better name.

> My grandmother, then a young girl of 15 or 16, had been reading Charles Dickens' *Barnaby Rudge* in which there appears a

character named Dolly Varden; also the vogue in fashion for women at that time (middle 1870s) was called "Dolly Varden," a dress of sheer figured muslin worn over a bright-colored petticoat. My grandmother had just gotten a new dress in that style and the red-spotted trout reminded her of her printed dress.

She suggested to the men looking down at the trout, "Why not call them 'Dolly Varden'?" They thought it a very appropriate name and the guests that summer returned to their homes (many in the San Francisco Bay area) calling the trout by this new name.

"That's kind of amazing," says Fiona. "A girl just like me named a fish."

"Well, you guys have named things too, you know," I say.

"Like what?"

"Like Otter Cove," I say.

Everyone chimes in.

"Like the Jumping Rock."

"And the Indian Trail."

"And the Diving Board."

"And the Lady and the Lamb."

"Right, these are things that are not on any official map but we know where they are."

Why do we name things? It gives us a shorthand reference. Like our own family language. We can let others in on the secret but only if we want to. On a broader level, we are putting our stamp on the place. We own it. It's ours to name.

I like that the kids name places. "Shelbriona Fort." Only Bronwen, Fiona and I know where this is. Actually, I think Michael followed us one day, so he might be in on the secret.

After dinner, we get ready for the campfire.

"I'll bring the marshmallows," says Nora.

"Fin, get your Swiss Army knife," says Roger.

Fin has his head in his book. Roger sighs. "Fin!" he shouts.

"Huh?" Fin closes his book and smiles up at his cousin.

"Let's cut some branches for roasting marshmallows," Roger says.

"Okay."

Darkness descends quickly this far north. Despite the heat of the day, the night is cool. I put on a fleece jacket, and go down to the beach where Kevin has the fire going. Earlier in the summer, he dug a fire pit and lined it with river rocks. He and Michael have rolled larger logs around the pit for us to sit on. Kevin likes to set up the campfire on the beach. He also likes to dig up grass and saplings that spread across the beach if they are not kept in check. Marking our territory.

Patrick carries down a couple of lawn chairs. Molly and Joanna gather throw rugs. I bring the guitar and songbooks. Kevin's New Year's resolution was to put together a Chilliwack Lake Song Book, and he has finished it. He and I exchanged song ideas through the winter. I put forward my Dad's old-timer songs. He added songs from his summer as a lifeguard at Pioneer Pacific Camp off the east coast of Vancouver Island, "Kumbaya," "We Are Climbing Jacob's Ladder." I add a few Peter, Paul and Mary favourites: "If I Had a Hammer," "Where Have All the Flowers Gone," Leaving on a Jet Plane." Kevin adds John Denver's "Take Me Home, Country Roads." By the time he's finished, he has seventy-five songs. He's printed out two books with chords for the guitar players, Kevin and me.

Once we're settled around the fire, the marshmallows come out. The kids jostle for the best position.

"I like my marshmallow brown and bubbly on the outside," says Nora.

"No, burnt is best," says Will.

Kevin and I tune up our guitars. We start with "I've Been Working on the Railroad," then "Quartermaster's Store." The songs remind me of Dad. He loved a campfire. The stars come out one by one as the night darkens. No one wants to go in. Will puts his stick in the fire and swirls the glowing end around his head in circles. A swirl of yellow against the darkening sky. I strum the cords to "Good Night Ladies," and the kids chorus.

"No. Not time to go in!"

Eventually, they gather up songbooks and blankets.

"Did you hear that?" Joanna says.

We stop talking.

"Hoot, hoot, hoot, hoot," or as Peterson would say, "Who cooks for you?"

The signature four hoots of the barred owl. Joanna, Bronwen and I get up and walk into the forest following the sound. Silhouetted against the sky, on a low branch, is the owl. Its round face swivels towards us. Then it lifts its wings and glides past us into the forest.

I learned while researching Stó:lō tales that the owl is the messenger of the spirits, bringing the wisdom of the Elders who have gone before. I wonder what kind of message our owl brings us, what wisdom from the past, what prescience of the future.

American Fishing Club

My parents and grandparents always talked about Charlie Lindeman as if he were family. Tales of Charlie are part of our family folklore. In 1942, he suffered a heart attack and moved to Sardis to live out his last months looked after by his good friend, Carl Wilson. Charlie died early the next year, and for another six months after that, his cabin lay abandoned.

When I do a land titles search, I discover that Ronald Stockton, my grandfather's lawyer and the man who negotiated the purchase of Poppy's land, bought Charlie's land from him before he died. I'm not sure why, exactly, unless Charlie needed the money. Stockton paid one hundred dollars for Charlie's 14.8 acres. Did Poppy orchestrate this purchase to help Charlie out? I'd like to think that Poppy looked out for Charlie in his final years.

In July 1943, Stockton sold Charlie's property to Walter Gilbert, a local pilot, for a nominal fee of one dollar. It was Walter Gilbert who charted a new direction for Charlie's property, one that Charlie might just have approved of.

I have heard this name before. Poppy used to talk about the pilot Gilbert. I think John MacLeod mentioned him when I met with him two weeks ago. I look back at my notes. Yes, here it is. John recalled that when he'd walked in to the lake in the spring of 1948, he found Walter and Jeanie Gilbert living at Charlie's cabin. They told John they were operating a fishing lodge and that they were calling it Lindeman Lodge.

I find a citation on the Canadian Aviation Hall of Fame website. Walter Edwin Gilbert was born on March 8, 1899, in Cardinal, Ontario. At eighteen, he joined the Royal Flying Corps. He trained at Camp Borden, Ontario, was commissioned as a flying officer in the Royal Canadian Air Force and served oversees during the Great War. After the war, he flew on forestry patrols in northern Manitoba. I find a photograph online. He was a handsome man, with strong features, hair short and slicked back. With his leather aviator's jacket lined with fur around the collar, he looked like Errol Flynn.

During the 1920s, Walter flew fisheries patrol on the British Columbia coast and freighted supplies to gold fields in Alaska. In 1930, the government of Canada asked him to join an exploration expedition to King William Island in search of the remains of Sir John Franklin's expedition to find the Northwest

Passage in 1845. During the expedition, Walter found a cairn containing artifacts of the Franklin Expedition. He talks about it in a book he co-authored with Kathleen Shackleton, *Arctic Pilot: Life and Work on Northern Canadian Air Routes*:

> There were just the pitiful remains of the rude canvas and timber camps which the stranded explorers had tried to erect to shelter themselves from the rigour of the Arctic climate—bits of iron, bits of wood, and torn remnants of the rotted canvas. Half buried beneath a cairn of rocks, a glint of faded blue.

Walter flew with fellow pilot Earl MacLeod into Chilliwack Lake in 1939, the year after the Big Fire. He befriended Charlie Lindeman and when Charlie became ill, Walter kept an eye on his cabin for him and visited him in Sardis, bringing him updates.

The year after Charlie died, Walter Gilbert and his wife Jeanie bought Charlie's property. The July 23, 1947 edition of the *Chilliwack Progress* reports:

> Walter Gilbert, a well-known former bush pilot and Western Airlines executive, has purchased the former Charlie Lindeman estate at Chilliwack Lake it is reported, and is moving in equipment and furnishings.

The couple converted Charlie's cabin into a dining room and built a kitchen on the back. They built two sleeping cabins on either side of the main cabin. They added homey touches, gingham curtains on the windows, washbasins on the commodes.

Walter flew guests in by plane. He installed a radio-telephone so guests could keep in touch with the outside world. The *Chilliwack Progress* did a story on the new venture at the lake:

> Something new in the way of a vacation land is being offered holiday seekers at Chilliwack Lake. Walter Gilbert, veteran flier, has turned to the tourist trade and the result is a fisherman's paradise, tucked away in the north end of the lake. Named after a veteran trapper who is one of the original lake settlers, the new vacation camp offers the kind of fishing that anglers dream about.

They nurtured their dream for eight years. However, Lindeman Lodge was not a financial success. Finally, Walter and Jeanie gave up. They moved to Seattle, leaving Lindeman Lodge vacant for several years. Charlie's property was abandoned for the second time.

I love the idea that Charlie's name stayed alive in the lodge. For my mother, Charlie was the heart of the lake. Although I never met him, I feel an emotional connection to him. I feel a stab of disappointment when I discover that the venture failed. I take some solace in the fact that his memory was kept alive. As I discovered from my research, in 1951, the lake which my mother and Charlie stocked with fish when she was a girl was renamed Lindeman Lake.

There is a gap in the story here. Years pass. Then in 1954 the place is sold to a group of American fishermen from California. Here is where the story goes cold. I search newspapers, archives, Ancestry.ca, Ancestry.com. All in vain. I can't seem to track down the connection between Walter Gilbert and the California fishermen.

How did these Americans hear about Chilliwack Lake, remote and little known even in Vancouver? I search the archives of the *Chilliwack Progress* where I discover a weekly fishing status report submitted by the Chilliwack District Game Warden extolling the good fishing at Chilliwack Lake in 1952 and 1953.

In the "very good category" is remote Chilliwack Lake where the fish are co-operating heartily with fishermen using both fly and troll.

Did they somehow track fishing news in British Columbia? Had they ever visited the lake? Was one of them at some point a guest at Lindeman Lodge? Did one or more have a business connection with Walter Gilbert? I find no answers to my questions. I only know that a group of fishermen from California bought Lindeman Lodge.

I thumb through Poppy's correspondence and find a letter from C. Ray Robinson of Merced, California, written on the letterhead, "Chilliwack Lake Lodge Ltd." I do a search of the British Columbia Companies Registry. Chilliwack Lake Lodge Ltd. was registered on May 25, 1954. All eleven directors lived in California, except for a Vancouver lawyer, Elmore Meredith. I start with C. Ray Robinson, who from Poppy's correspondence seems to be the spokesperson for the group. I search online and in local newspapers; I contact the Museum Archives in Merced, California. I find Robinson's death certificate and that of his only son. I try the other directors. Nothing.

The purchase of the property by the group brought about immediate changes. That first summer, they hired Ray Wells as manager to run the lodge from May to September. Ray's wife Alma remained in Sardis. In his diary, Ray writes:

I went in to Chilliwack Lake to run a private resort lodge for an

American company. I managed this lodge for the past nine years, and when this job finishes, I am going to take (brother) Ron with me and we will open the best Hunting and Fishing Camp in the Yukon. Be seeing you.

Starting in 1955, new boats were brought in, the wharf was upgraded, a new radio-telephone installed and a new games lodge built. The fishermen arrived by float planes through the summer months. My parents and grandparents resented the disturbance. During the years that Charlie's cabin was abandoned, my family had had many summers of complete isolation. Now planes were zooming in on a weekly basis and fishing boats could be seen on the lake every day.

The co-owners of the old Lindeman Lodge called themselves the American Fishing Club. The same men came year after year, sometimes bringing friends. But neither Poppy nor any other member of my family got to know them. They were interlopers. Private fishing clubs in British Columbia sprang up in the 1920s. The Pennask Lake Fishing and Game Club, established in 1929 by Hawaiian pineapple magnate James Drummond Dole, created an exclusive fishing experience for its members. The club bought all of the land surrounding Pennask Lake except a small reserve that has since become a provincial park. Luckily, the American Fishing Club had only a toehold on Chilliwack Lake, and their tenure was short-lived.

Bronwen, Fin and Will are asleep in the bunk room. It's eight in the morning, my time for quiet contemplation. I mull over what I have learned this summer so far. My quest is like unwrapping a box only to find another box inside, and another, until finally the treasure is unveiled. However, sometimes I feel that, somehow, I never quite get to the prize.

Almost all vestiges of the American Fishing Club are now gone. Last night, I drew a plan of the property with the buildings and paths, as best I could remember them. I want to take the kids to the provincial park to see what we can find.

I lean back in my deck chair on end of the dock. Murphy sits by my side. The sun arrives, a morning caress. I gaze across the limpid water, stillness washing over me. Minnows break the water's surface in droplet swirls. The sandpiper on the shore bobs its head. I listen to the sough of wind over water.

"Morning, Rani," shouts Bronwen from the porch.

I wave back. Murphy races back to greet her. I gather my cup and chair and walk across the beach. The sand prickles the soles of my feet. I'm excited to get the day going, to share my map with the kids. I savour every minute I have alone with Bronwen, Fin and Will. In no time, they'll be grown, working at summer jobs, no longer able to spend the month of August with their grandmother at the lake. When I come in the sliding door, Fin is reading in the corner chair, still in his pyjamas.

"Hey, Fin," I say. "Let's have some breakfast."

"Okay, Rani." The voracious reader gives me a smile, puts his book away and goes upstairs to get dressed.

Will pulls over the stool and retrieves the cereal from the top shelf. We sit outside at the picnic table on the porch with our bowls of cereal, blueberries, orange juice and toast. Over breakfast, I show them the map and talk about the fishing club that was here when I was a girl.

"The club was where the provincial campground is now, right?" asks Bronwen.

"Yes, it started when I was very young. They bought the land and cabin that used to belong to Charlie Lindeman, and he died before I was born. So the fishing club was there as far back as I can remember.

"The fishing club had a wharf and boathouse at the beach. The boathouse was where the provincial campground boat launch is now. A trail led from the beach to the top of the bluff where a dining lodge and kitchen, a games lodge, and five sleeping cabins sat in a natural clearing. Seven buildings in all. Each cabin could sleep two guests. Ray Wells was the caretaker and was assisted by a cook and chambermaid. Ray became a good friend of our family. He was over for dinner all the time."

I pause to point out a cabin on my hand-drawn map.

"One of the sleeping cabins was set on stilts at the shore's edge down a steep trail from the top of the bluff. When no one was around, I'd sneak down and sit on the porch overlooking the water."

"Did you ever get caught?" asks Will.

"Nope."

"Why'd you go?" asks Bronwen.

"I liked being alone. I'd go in the afternoon, when the chores were done. My grandparents were sleeping. No one knew I was there."

"I'd like that too," says Fin. He's the most like me that way. He is quiet, watches and listens, but keeps his thoughts to himself.

"Why aren't any buildings left?" asks Bronwen, bringing us back to our morning's outing.

"The fishing club was sold to the government for a provincial campground. The buildings were abandoned. Over the years, they were broken into and eventually burned to the ground."

"Who would do that?" asks Fin.

"I don't know," I say, "I was away in Zambia at the time. But I was really upset when I heard about it."

"I would be too," says Will.

"Let's go and see what remnants of the Fishing Club we can find," I say.

After breakfast, Fin does the dishes, Will fills the wood box and Bronwen tidies up the living room, putting the cards and games away from last night. I put a few large plastic bags in my pocket for Will, always on the lookout for garbage. We walk in front of Cupola Lodge and around the group of rocks in front of the Sandpiper and continue down the half-mile of beach until we are in front of the MacLeod's cabin. We wave to Flora's family who are here for the week. Next to the MacLeod's is the gravel road that comes down the hill and ends in the provincial campground boat launch.

"A nice beach used to be here. White sand, just like ours," I say.

"Too bad it was wrecked by the boat launch," says Fin.

"Well, if there was no boat launch, we couldn't get our boat in," I say.

"Hmm," says Fin. "You're right."

I remember Mother railing against the desecration of that beach. I didn't share her outrage. It's not the beach we swim at. Ours is a kilometre away. We use that boat launch to put our motorboat in every spring and take it out every fall. So it's actually pretty useful to us. But I see her point. She remembered that beach when no one else was here except Charlie Lindeman.

We take the path up the hill to the bluff overlooking the lake.

"The fishing club was at the top of this path. "

"See if you can see anything, look up in the trees."

"Hey, I see something," says Fin. "A piece of cable."

"The fishing club had a wooden cart that carried supplies up this path on a rail," I say. "The cable was connected to it."

"Look," says Bronwen. "There's cable high up on that tree."

We walk across the clearing at the top of the bluff and look up. We can see rusted cable near the top of the tallest Douglas fir. A few wooden slats are nailed halfway up the tree.

"That's the remains of a ladder to a radio antenna on the top of the tree," I say. "Look over to the left."

I point to a crumpled heap of stones and mortar streaked by lichen. The bones of the fireplace of the games lodge, all that remains of the seven fishing club buildings.

Gran asks me to row over to the fishing club and invite Ray Wells, the manager, for dinner. Now that I am eight, I am allowed to row by myself.

I step out of the rowboat with a smile, eager for Ray's welcoming eyes and easy-going manner, but he isn't down by the water.

I pull the rowboat up on the beach beside the boathouse, its doors gaping open onto the shore. The wharf in front of the boathouse is lined with fishing boats. The next group of guests doesn't come till next week so all is quiet. I walk across to the wood steps that lead up to the bluff. Beside the steps, a wooden cart sits on a rail track, waiting to transport supplies from the beach up to the bluff. The cart is pulled up the track by a winch operated by a small gas engine at the top.

A few weeks ago, my brother Chris decided to run the cart by himself. He cranked the engine, opened the throttle and got in the cart. It careened down the rail and flew off the end of the track splitting into pieces on the rocks below. He got a bump on his head and a dressing down from both Ray and Poppy. He had to help Ray build a new cart. Chris is braver than me. Even if this idea had occurred to me, I wouldn't have had the nerve to try.

At the top of the path, the cabins for the fishing club are arranged around the clearing. Front and centre is the dining lodge, originally Charlie's cabin. I will likely find Ray here. I climb the steps leading up to the porch, past flower boxes filled with geraniums and daisies.

"Hello?" I call out. "Ray?"

The dining room has a small wood stove at the back of the room. Red and white gingham curtains frame the windows. An oilcloth tablecloth, white with yellow flowers, is tacked in place over the edges of the dining table. At each end of the table is a coal oil lamp.

"Come into the kitchen," Bertha calls out.

I go out back into the kitchen. Bertha is the cook, a jovial, full-bosomed woman from the valley. I am in luck. She has a vat of oil on the wood stove and is mixing the batter for doughnuts.

"Like a doughnut?" she says.

I smile and nod my head. She drops a couple of spoonsful of batter into the oil. Sizzling and popping, the batter rises and browns. Using an open-slatted spoon, Bertha scoops up the doughnuts and drops them into a bowl of sugared cinnamon. As soon as they cool, she hands one to me. I take a bite, a crispy outer crust, sugar melting in my mouth, spongy inside.

"Mmm, delicious," I mumble, mouth stuffed full.

"Ray is in the lodge fixing the fireplace draft."

I walk over to the games lodge, admiring the large windows looking out over the lake and stone fireplace at one end. It sparkles like a shiny penny compared to our rough-hewn grey weathered cabin. I call out to Ray as I come to the door.

"Come on in," Ray shouts.

Ray sits on the fireplace hearth, tools in hand. I take note of the pool table with its green baize top, chairs covered in maroon leather placed in front of the fireplace. In two of the corners are round tables and chairs for card games.

"Gran wants to know if you'd like to come for dinner tonight," I say to Ray.

"Well," he says in his usual slow drawl, "that would be mighty fine."

I look up at Ray. Large-boned, broad-shouldered, and big-bellied. Ray is a man in his sixties, yet he can buck up a tree and sling rounds of wood as if they were matchsticks. He has a wide forehead and pale blond hair, sparse, swept back and slicked down. His eyes greet me with amusement, like he has a joke to share. His nose looks like it has been broken more than once. I have never seen him without a wad of chewing tobacco in his cheek. It makes him sound like his mouth is full of marbles when he talks. I listen hard when Ray speaks, not wanting to miss the pearls in his mumbled speech. He has a habit of sucking in his lips to make a faint clicking sound. He often has a cigar clamped between his teeth, not necessarily lit, just there, as if it is part of his face.

I sit on the edge of the hearth and watch while Ray fiddles with the draft.

"Goldarn thing is rusted out," Ray says. "I'll need to make a new one. Come on with me to the tool shed. Let's see what we can find."

I happily follow Ray to the shed behind the woodpile. He whistles as he looks for the right piece of metal. On the haphazard shelves are jars full of nails and screws, bits and pieces of tin and wire. He shows me a wooden water wheel.

"This was made by old Charlie. He put it in the crick out yonder and the water turned it to make power. I'm keeping it for old times' sake," he says.

I turn it over in my hands. I like the way the wood pieces fit together.

"Oh!" I say, suddenly remembering that I'm supposed to come right home. "I better get back. Mother will wonder why I'm gone so long," I'm almost out the door when I turn one last time. "Oh, and Gran says six o'clock."

"I'll be there," Ray smiles.

Ray arrives on the dot of six. He has on his brown felt cowboy hat. He removes it with a flourish when he opens the door and sees Gran.

"Much obliged for the invite," he says.

Ray's hair is slicked back. He has on his best plaid shirt with a bolo tie, braided leather held by a silver clasp. He is a keen admirer of Gran's berry pies, "the best pies this side of the Rocky Mountains," he likes to say. Ray has a story for every circumstance. Chris and I love them. Ray has a country twang and colourful expressions.

He starts most of his tales with "I've heard it said," then off he goes. He keeps us entertained throughout the meal.

I remember a story about Jack Post, a prospector living in the Chilliwack Valley, and a friend of Charlie Lindeman's. Ray tells this story on the tape I borrowed from the BC Archives.

> One night Charlie was coming down the lake from Brown Crick, paddlin' down the middle of the lake, and Jack was paddlin' along the shoreline and he hollered at Charlie.
> "Is that you, Charlie?"
> "Yes, Jack."
> Jack says, "What month is it, and what year?"

Ray laughs so hard on the tape that it's a minute before he can continue. "He wasn't too interested in what time or what day it was, just what month and what year."

Another story he told was about Ed Allison, who ran packhorses up to the Red Mountain Boundary Mine in the early 1900s.

> Old Ed, he was up at the Boundary Mine. He was commenting on how he liked bakin' powder biscuits. But he didn't have a stove to bake 'em. So the mine manager gave him an old camp stove.
> After dinner, a couple of men helped Ed put the stove on a packboard, and helped him get it onto his back. When he got home, why, he set this stove up and started mixing up a batch of bakin' powder biscuits. When he opened the stove, he finds the oven full of green alder wood.

Ray starts to laugh. He has a hard time finishing this story, too. "I guess whoever had used that stove, had put the wood in the stove to dry. Old Ed had quite the load to carry that day."

After dinner, we sit around the dining table and play dice games. Ray throws the dice. He doesn't much like what he sees and mutters, "What a bunch of coconuts."

He chuckles to himself. I catch Chris' eye and we grin at each other. We both like that expression. Chris uses it when there's no one else around but me.

Ray was already in his sixties when I knew him. What was his early life like? How did he end up as manager of the American Fishing Club?

At the Chilliwack Archives, I find *Edenbank: The History of a Canadian Pioneer Farm*, edited by Marie Weeden. It's the story of Ray's family, among the first settlers in Sardis. I look up Marie and find her in the phone book and give her a call. She now lives in a residential development called Edenbank, built on the original Wells farm. She is Ray's niece and can tell me lots about him. She considers herself the Wells' family historian. I arrange to visit at the end of the week, on a day that Kevin and Joanna will be at the cabin so they can keep an eye on Bronwen, Fin and Will.

Our cabin is fifty kilometres from Marie's house, so it's a day's outing. Marie and her husband, Dick, greet me at the door. In anticipation of my visit, Marie has spread boxes of papers and photographs across the dining room table. She offers me tea and ginger cookies and we spend the afternoon looking through the boxes for photos of Ray. At seventy-eight, she is spry and has answers to my questions at her fingertips. She shows me a copy of the Wells family tree and explains the family relationships. Most of the photos of Ray are of him on horseback, a place he obviously feels very comfortable.

"Well," Marie says, "Ray was an unusual man, kind of difficult and cantankerous at times."

We pore over photographs, mostly of men on horses, trying to recognize Ray. She does better on the younger photos; I do better on the older ones. She shows me a few yellowed newspaper articles. In "A Joyride Back in Time," the columnist, Bruce McLean, takes a ride with Ray in his horse and buggy down the streets of Sardis. Here's Ray, cigar hanging out of his mouth, Stetson on his head, blanket across his knees and a big smile on his face. Marie and I smile as we look at this photo together.

"I loved his stories," I say. "Whenever he talked, he put a spell on my brother and me as kids."

"What a bunch of coconuts!" says Marie, with a laugh.

"Oh boy," I say. "That's Ray in a nutshell!"

In a *Chilliwack Progress* article, Ray is described on the occasion of his eightieth birthday as "one of the last of the old-time gentlemen." I think to myself, he sure was.

"Do you have any photos of the fishing club," I ask. I have been on the lookout but haven't seen any.

"No," she says. "I wish I did."

I am crushed. I have this image in my head of the lodge and sleeping cabins and desperately want to compare it to the real thing. I can't believe

Ray Wells, grandson of early pioneer Allen Casey Wells, seated on a painted pony, circa 1904. Ray became the manager of the fishing lodge at Chilliwack Lake in 1954. Courtesy of the Chilliwack Museum and Archives, 2004.046.003.

how parts of history are just lost. I am frustrated, feeling like there must be photos somewhere. I will have to redouble my efforts. With the reach of social media, maybe something would turn up.

When I finally look up from the piles of papers, I realize I have been here for hours. I apologize for taking up so much of her time, but she is eager to talk to someone interested in her family history. Her husband Dick retreats to the study but he might as well be right beside us.

Marie calls out to Dick, "Do you remember that time Ray flew into the lake with the new fishing boat? And it got loose and crashed into the trees?"

"Where is that photo of Ray with his packhorses fording Post Creek?"

"Can you photocopy this letter for Shelley?"

Dick graciously obliges. They are both so accommodating. They are welcoming me into the circle of historians, the preservers of the past.

Just as I am leaving, I ask Marie whether she has anything that Ray might have written.

"Oh yes," she says. "My grandfather wanted all the boys to keep a diary to record important life events. Ray was reluctant, but he did write down a couple of pages. I typed them up. I have them somewhere. Dick, do you know where Ray's diary is?"

Dick finds the six typewritten pages, *An Autobiography—by Ray Wells, January 1962* and makes me a photocopy. I say goodbye, thanking them both for their help. I don't wait until I get home. I sit in my car in the dwindling light and read Ray's diary. After opening all the boxes, I feel today like I have unveiled the prize. Ray springs to life on the pages.

He starts at the beginning. In 1865, his grandfather, Allen Casey Wells, travelled by paddlewheeler from Yale to Chilliwack Landing, a stopover station for miners on their way to the gold fields on the Fraser River. Allen thought the land in Sardis looked promising for a dairy farm. He bought three hundred and twenty acres of land on the Luckakuck River, a tributary of the Chilliwack. There were no roads back then, only lakes, rivers, and occasionally, a trail through the forest.

Over the next few years, more settlers arrived. Wagon trails were cleared, and a school and church built. The Sunday church service provided a community gathering place for both Indigenous and white people to sing and worship together. Getting to church was a challenge with several streams to cross between the church and the Wells farm. Allen kept a canoe cached at the first stream. He paddled while his wife, Sarah, held a line to their horse that swam behind the canoe. Reaching dry ground, Allen dragged the canoe, tied by a rope to the horn of his saddle, to the next stream. After several such crossings, they reached the church. Going to church was the main social event of the week. Everybody would visit before and after the service, exchanging gossip.

Allen's house in the wilderness gradually became Edenbank Farm, a model of progressive dairy farming. The construction of the Canadian Pacific Railway through the Fraser Valley provided access to markets across Canada for dairy products. Allen was constantly applying new ways to improve his dairy herd. He was described in the *Chilliwack Progress* as "the father of dairying in BC."

Allen's son, Edwin, inherited his father's zeal for innovation. In 1893, Edwin married Gertrude Kipp, another pioneer family in the Chilliwack Valley. The following year, Ray, the first of seven boys, was born in the house that his grandfather had built.

One of the earliest farms in the Chilliwack area, established by Allen Casey Wells, circa 1870, along the Luckakuck River. Edenbank Farm became a model of progressive dairy farming. Courtesy of the Chilliwack Museum and Archives, P3542.

Of course, Ray told the story in his own fantastic way.

On July 31st, I floated into Edenbank Farm on a drift log from Paradise Valley. Mother said I was the cutest baby there ever was. I weighed ten pounds and everything worked. According to Mother's diary, I was very slow to learn and to talk—and I've remained that way all my life.

The Wells family had a strong link to the Indigenous community. Chief Sepass, who worked at Edenbank Farm for Ray's grandfather and for his father, became a role model for Ray, giving him a glimpse of the world beyond the farm. Chief Sepass taught Ray to shoot, took him hunting and trapping.

When Ray was eight, his two younger brothers died of diphtheria, a day apart. When he was twelve, the hogs on the farm contracted a disease. Ray's job was to drive the hogs into a large pit, shoot them, and cover them over in lime and dirt to stamp out the disease.

At the start of the First World War, discontent erupted at Edenbank. Haying season came to a halt. The men refused to work. Ray, aged twenty-one, strong and burly, marched into the bunkhouse and took on the ringleaders. Work resumed the next day.

Ray attended agricultural college and went to work on the 76 Ranch in Crane Lake, Saskatchewan. When he was twenty-four, Ray married Alma Campbell, also from the valley. "Up to then," Ray wrote in his diary, "that was the wisest thing I had ever done, and the next year she accompanied me back to the 76 Ranch again. She enjoyed it there as much as I did."

Ray and Alma moved from Saskatchewan to a ranch in the Kamloops region of British Columbia. Ray's diary describes the birth of his only child, Lloyd, in 1919:

> The day Lloyd was born—I'll never forget that day—all hell broke loose because he arrived about two months ahead of schedule. I ran a saddle horse three miles and back to phone Kamloops for a doctor sixty miles away, and to get a neighbour lady, Mrs. Foster, to come and be with Alma. The Doc arrived about an hour too late but said everything was OK.
>
> I remember he only charged $80 because Lloyd only weighed 3 lb. 14 oz. The clothes Alma had were all too big so they wrapped him up in cotton batten and put him in a sewing basket. The Indian woman that used to do the washing and scrubbing for Alma made him a birch bark canoe for a crib. When he got going and grew too big for the canoe, she made him a larger one.
>
> Lloyd cried day and night for a week until we tried Eagle brand condensed milk. We raised him on that, and today at forty-two years, he is up to two hundred and fifty pounds. So much for Lloyd.

Ray was restless. He moved from the interior of British Columbia to Alberta and back to Saskatchewan. Finally, he and Alma decided to settle in Chilliwack and Ray worked at the Fraser Valley Utility Plant. A few years later, he took over the Cultus Lake Riding Stable.

Ray heard about an outfit out of California that had bought old Charlie Lindeman's place up at Chilliwack Lake. He approached the spokesperson of the group and put himself forward to manage the fishing club. From what I learned from Marie Weeden, this move fit Ray's personality. He was un-settled, switching jobs often. He was an independent spirit, but he liked a challenge and at sixty, he was still physically strong, with lots of good years left. He was a jack-of-all-trades and was familiar with Chilliwack Lake, hav-ing ridden in there often. The remote location appealed to him. He fit the bill perfectly, an eccentric old-style storyteller. He was forged in the same mould as Charlie.

"The Americans are ruining the fishing," Poppy said to anyone who would listen.

As a girl of six, I was excited with the newness of the change. I didn't recognize the disturbance to the peace and quiet that our family had come to take for granted. Although we were told to be polite, our family kept their distance from the fishermen. Chris and I were only allowed to visit Ray when no guests were around. And I was supposed to row straight back home if I was sent to deliver a message to Ray. I think Mother was worried that once the men started drinking at the end of their day of fishing, things could get out of hand, or at least, beyond what she was comfortable for me to see.

I am eleven. Dad and I are at the fishing club. It is late afternoon, nearly dusk. We are getting ready to row back to our cabin when a squall comes up, sudden and fierce. A driving wind crashes waves against the beach. The club's fishing boats are banging so heavily against the wharf that some break their moorings.

Dad tells me to stay high on the beach away from the waves. He sprints up the trail to the lodge shouting, "Come quick Ray, the boats are being smashed."

Ray comes at a run. He's in his mid-sixties, but he's still strong and agile. He and Dad jump into the water between the boats and the wharf. I'm frightened, too small to be of help, but old enough to know how dangerous this is. The two men wrestle the boats to shore and pull them up onto the beach to safety. The wind dies down as quickly as it began. No one comes to help. Where are the other men? I don't know, but, in my eleven-year-old mind, I blame them for not lending a hand.

We carry on with our summer life but the fishing club intrudes. The sound of an airplane is like the ring of a telephone interrupting our solitude.

I sit on the porch, Poppy teaching me how to play cribbage, and listen to the soft droning noise, getting louder and louder. I run out onto the beach and, eventually, I see a plane coming from the valley.

"Where's it headed?" asks Poppy.

Starting as a speck in the sky, once it is directly overhead, the plane looms large, pontoons clearly visible. It flies straight down the lake before turning and heading back towards me. I lose sight of the plane against the dark green of the mountains but spy the splash of the pontoons as they set down on the water's surface.

Will the plane veer towards our end of the beach or go to the fishing club? If it veers towards our end, it will summon Dad back to the office, an early end to his summer holidays. I don't want him to go.

"It's going towards the fishing club," I call out, relieved.

"More fishermen," Poppy grumbles.

The Americans came and went through the summers of my childhood. The visits became more sporadic as I entered my teens, then they stopped altogether. Every so often Poppy received letters from Mr. Robinson saying that Charlie's old property was for sale. I discover that, in 1964, my uncle Rowly made an offer of thirty thousand dollars. Mr. Robinson said he'd put the offer to the members, but the communication trail ends here. Why would Rowly want to buy Charlie's? He had hated his winter with Charlie and here he was trying to buy a piece of it. Maybe he loved the lake more than we knew. Maybe he wanted his own place away from the constraints of the family. I wonder what would have happened if his offer had been accepted. It certainly would have rewritten our family history.

In September 1965, Mr. Robinson wrote to tell Poppy that the group had decided to close down the fishing club. No one, he wrote, was allowed to use the buildings without his written authorization, including Ray Wells. I feel an affront reading this. How dare he imply that Ray was not to be trusted!

After the fishing club closed, Ray returned to Sardis. He lived into his nineties, forever an independent soul and a character. He had no time for the new-fangled motorcar and he suffered from narcolepsy: he could often be seen driving the streets of Sardis in a horse and buggy at all hours of the day.

I was seventeen when the fishing club closed, just finishing high school. The exchange of letters around the closing of the fishing club meant nothing to me. It was just background noise to my own exciting life. I was off to university, and for the next four years, spent little time at the lake. I vaguely remembered my parents discussing buying the fishing club property. They decided that they didn't want to take on more land. I often wonder now how things would have evolved if they'd bought Charlie's land. Perhaps Patrick and I would have built our cabin there. It's a long way from the rest of our family's cabins, not within easy walking distance, and maybe a ten-minute paddle. I wonder if the MacLeods wish that they had bought it: they would have had room to expand. Today, I think how amazing it would be to have the whole lake to ourselves, just us and the MacLeods.

In the early 1970s, there was a change of government in British Columbia. The New Democratic Party came into power. One of the New

Ray Wells driving the streets of Chilliwack in his horse and buggy, circa 1990. Ray was a figure around town with his trademark felt hat and cigar clamped between his teeth. Courtesy of the Chilliwack Museum and Archives, 2005.047.001, File 3.

Democratic Party's priorities was creating more provincial parks. Poppy and Earl MacLeod received letters from the government inquiring whether their properties at Chilliwack Lake were for sale. Neither family wanted to sell. But the California club members did. The government bought the land where Charlie built the first cabin on the lake for seventy-five thousand dollars. Two years later, it became the Chilliwack Lake Provincial Park.

The prospect of a public presence filled my parents and grandparents with dread. A campground opened with a handful of camping spots. A bulldozer dug a swathe down the hill to create an access road to the beach. Rocks were shoved onto the beach, ruining the stretch of white sandy beach. Gravel was brought in and a boat launch created. Mother mourned the destruction the most keenly. By then, my grandparents were in their eighties and rarely at the lake.

Patrick and I were living in Zambia when all this was taking place. We returned in August 1972 to find the fishing club vandalized, the cabins reduced to burnt-out husks, with only the fireplace chimney of the games lodge still standing. Mother railed at their destruction. I was mournful rather than angry, part of my childhood disappearing in what seemed to me a blink of the eye.

NEIGHBOURS

In early September of 1982, Dad saw a rusty red Ford parked on the side of the forestry road near our access road turnoff. He pulled up beside the truck where a man was loading up his provisions. He was in his late fifties, with a full beard, greying hair, and weathered face. He answered in monosyllables, wary of Dad's questions. He filled canvas provision bags with coffee, sugar, tins of corned beef, flour and oatmeal. He strapped the bags onto a packboard with a leather harness and a head strap.

"You live up here?" my Dad asked.

"Flora Lake," came the response.

Flora Lake, named for Flora MacLeod, the matriarch of the MacLeod family, is a small subalpine lake over the ridge east of Chilliwack Lake that sits at 1,700 metres in a dip 300 metres below the ridge.

When Dad told us the story, he called him The Old Trapper. The name stuck, the way names at the lake seem to.

The Old Trapper had cut a trail beside Cupola Creek up to the ridge and down the bluff on the other side to Flora Lake. He felled some small pine trees and built a cabin. He carried in his supplies on his packboard. During that fall, he and Dad saw each other every few weeks.

"Can I park my truck inside your gate?" the Old Trapper asked Dad one afternoon in late October. "I want to keep it away from prying eyes."

"Sure," said Dad. "It'll be fine there."

Over the next several summers, we often saw the red truck tucked inside the gate. From time to time, Dad would run into the Old Trapper. He had finished his cabin and was living at Flora Lake full-time. Except for taciturn greetings, the Old Trapper never told Dad anything about himself or his past. I saw his truck a few times but I never met him. None of us knew his name. I wasn't sure about him. Was he dangerous or merely eccentric?

What had brought the Old Trapper to the lake?

In these last few days of summer, I think about him living out here all year, alone. I wonder if I can find out anything about this man. I ask around. A few remember him but no one can shed any light on who he was or where he came from. I ask Shannon at the Chilliwack Archives. No luck. The Old Trapper remains an enigma. All I can say is that he was a man in search of seclusion and he found it here. Just like Charlie Lindeman. Kind of like me; we share the need for solitude.

In 1984, Patrick and I decide to hike to Flora Lake on the Old Trapper's trail. We are still living in Calgary and are staying at Cupola Lodge with the boys for a three-week holiday. The boys, who are ten and twelve now, want to stay behind with their grandparents so we head off on our own for the day. We find the trailhead beside Cupola Creek just off the forestry road. The path is well maintained, easy to follow but not easy to climb. It's steep, with short switchbacks. Before long, my legs are burning. We stop at the edge of a narrow falls, shouting at each other to be heard over the noise of the cascading water. The forest gives way to alpine brush, with pockets of red paintbrush, blue penstemon and pink spreading phlox. We take in the view over Chilliwack Lake and the Cascade Mountains in the distance. We pause, soaking up the stillness. At the top of the ridge, we stop for lunch and a rest.

"Listen," says Patrick. "Did you hear that?"

I cock my head. Sure enough I hear a cheep, like a newborn chick but more emphatic.

"A pika," I say, pointing to the rocks on the ridge. "Look."

Pikas, the smallest of the rabbit family, pop up their heads and bark out alarm calls. Alert, their rabbit-like noses twitch in anticipation. Rounded dark grey ears edged with white stand out against light grey-brown fur. Winsome and comical.

"Let's see if they'll come closer if we toss some bread crumbs," says Patrick.

We cast a few on the rocks beside us. The pikas don't move.

"They'll gather them up after we're gone."

After lunch, we push ahead and crest the ridge. Standing on the ridge looking away from Chilliwack Lake and down the precipice, we can see Flora Lake below. We work our way down the bluff for another hour and soon realize that we will run out of daylight if we don't turn back. We won't make it to Flora Lake and we won't find the Old Trapper's cabin. Not today.

It is a gruelling return trip and my legs feel like rubber when, at dusk, we stumble onto the forestry road. How does the Old Trapper make this brutal trip with a packboard? And he's probably thirty years older than me!

A few years later, we head back up the Old Trapper's trail. By then, the path is overgrown and impassable. We give up on the idea of a return visit. I ask around and hear that the Old Trapper has disappeared. The cabin at Flora Lake is abandoned.

I wonder whether anything remains of the Old Trapper's cabin. We'll have to hike in sometime and see what we can find. Maybe later this summer.

How strange to think I had a neighbour I never once saw. That happens in apartment buildings in the centre of big cities, but not out here in the wilderness, where people need each other. We knew the Old Trapper existed, but we felt his presence only in the abstract. The Old Trapper lived a completely solitary life, only a day's hike from our cabin. I wonder if such a thing would be possible today.

Other neighbours we *did* get to know: the people who built cabins at Post Creek, a summer colony three kilometres downriver from Chilliwack Lake. The community was established sometime in the 1970s. Why here? I wonder. According to my faithful companion, *In the Arms of the Mountains*, the Post Creek Community was named after Jack Post, a gold prospector in the Chilliwack River valley in the 1890s who founded the Lone Jack Mine in the Cascade Mountains just across the United States border.

The push for recreational property started to build once the Chilliwack Lake Road was completed in 1961. Feeling the pressure, the Forest Service put forward a proposal to establish what amounted to a subdivision of summer homes at Post Creek.

The Chilliwack Regional District, a local level of government, opposed the Post Creek Community, noting the lack of adequate services:

> Anyone choosing to improve one of these lots is faced with a ground covering of solid boulders anywhere from three to eight feet in diameter, with absolutely no topsoil to be found, an entire surrounding neighbourhood of six to eight foot jackpines which will have difficulty growing any taller because of the soil conditions plus a mile or more walk to the lake which is too cold for swimming in, even during the hottest summers.

I smile at the idea that the water is too cold for swimming. We seem to manage even in the coldest summers. We hardy pioneers!

Despite these objections, the proposal went ahead. In 1970, the Ministry of Lands, Forests and Water Resources held a public auction for the lease of eighty-five lots. The lease payment was fifteen dollars plus the first year's rental, varying from twenty-three to seventy-five dollars per year, with an option to purchase the lot after a "habitable dwelling" was constructed. The road was not accessible all year round. There was no water. A request to bring in electricity was refused. Despite these obstacles, all eighty-five lots were taken up at the auction. The new owners were hardy and independent souls, many of whom cleared their land and built their cabins by hand. One homesteader painstakingly built his stone house over several years, selecting his stones from the river, mixing cement and layering stone and cement until the walls gradually took shape. Initially, we made no effort to get to know these neighbours. We passed the turnoff to the community a few kilometres before the provincial park and kept on driving.

Over time, Mother met a few cabin owners and joined the Post Creek Ratepayers Association. I went with Mother or Dad to the annual general meeting every May long weekend. I felt awkward, not wanting to voice my opinion for fear of interfering. I suspected that these folks thought we were snobbish with our swanky lakeside "mansions." But Patrick and I did make friends with a young couple who moved to Post Creek a decade or so after the community started.

It is the third week of August and Patrick and I are having a quiet morning of reading on the porch. Kevin and Joanna have taken the kids and Murphy on a hike for the day.

Two kayaks appear in the distance, ripples trailing in a V behind them. As they get closer, we recognize Dan and Erin Coulter and wave them in. I put on another pot of coffee and we pull two more chairs onto the porch. Dan is short and wiry, with buzz-cut greying hair and piercing blue eyes in a square face, hands rough from years of manual work. Erin has a direct gaze from deep brown eyes and is quick to laugh, her short-cut brown hair curling impishly around her face. We greet each other with the ease of a friendship built over many years.

Dan and Erin tell us about the parcel of land on the Chilliwack River ten kilometres upriver from Vedder Crossing that they bought last year. They planted garlic to sell in the farmers' market in Sardis. A pair of their goats had kids. Goat's cheese will be next for the market. They are off-grid, using in-river power generation from a nearby stream.

"We put in hydro a few years ago," Patrick says. "Do you want our solar panels? We're not using them now."

"For sure, that would be great," Dan says. "Let's do a trade, logs for solar panels. I can see a couple of your porch railings need replacing."

I tell Dan and Erin about the research I am doing on the Post Creek Community.

"Tell us about your early years," I say.

"Well," Dan begins. "I bought one of the lots in the mid-seventies. I started building a cabin. I could only work on weekends, so I didn't make much progress."

"What made you come here in the first place?" I ask.

"Well, my mum and dad had bought a lot there. I helped them build their cabin. I loved the peace and quiet. And I loved the challenge of doing everything myself."

In December 1982, Dan and Erin moved to the Post Creek Community full-time, with Erin's three-year old daughter, Megan. The cabin was still unfinished: no running water, no electricity, and no telephone. No flooring, kitchen cupboards or front stairs. Like Thoreau at Walden Pond, Dan and Erin embraced simple living and self-sufficiency. They were the only year-round residents at Post Creek. The last eight kilometres of the road was gravel and treacherous in the winter.

Their daughter Emily was born a few years later. Erin boiled Emily's diapers on the wood stove, adding a teaspoon of Ivory Snow, then carried them a kilometre to the river to be rinsed. Erin pushed Emily in a suspension baby carriage and carried the pail of diapers.

"I wore the rubber right off the tires of that buggy," Erin laughs.

Dan and Erin both worked part-time. Dan did construction work at Post Creek and in the valley, building porches and stairs, renovating cabins. Erin worked at Pioneer Building Supplies in Sardis, forty-five kilometres away.

They used a bucket system for water. Dan would fill ten five-gallon buckets each week in the river and line them up on the back porch.

"After a few years, I hooked up a gravity system. But the pipes froze in the winter. So we went back to the bucket system," says Dan.

"I remember when you first came to visit us at Post Creek," says Erin.

In the summer of 1985, Patrick and I move to Vancouver from Calgary. Our boys are thirteen and eleven. We hear that a young family is living at Post Creek. We want to meet our new neighbours, Dan and Erin Coulter. In early July, we drive the three kilometres from our cabin to the Post Creek Community. The trees are scrub Jack pine, two to three metres tall. The land is rocky and dry, inhospitable. The lots are identified with a wooden post with the lot number painted in black. Mother told us the lot number to look for.

We pull into their driveway, more of a track really, and see Erin tending her vegetable garden. We get out of the car and introduce ourselves. Dan has made four raised vegetable beds each about one metre wide and two metres long. The wooden boxes sit up off the ground just less than a metre.

"These beds are working really well," Erin says. "The soil warms up quickly in the spring and we get a good growing season. They make all the difference for vegetables."

The beds are enclosed with wire netting to keep out the deer. We watch as Erin weeds the squash and potatoes. She has staked the fledgling tomato plants.

"We don't use any pesticides," she tells me. "I plant mint and basil to keep away snails, and impatiens and verbena to deter rabbits. They seem to be able to get under the fences, even when we bury the netting."

Dan joins us with a pot of coffee and two extra mugs and we sit in the sunshine on the steps to their porch. Dan and Erin talk about their philosophy. We are in awe of their single-minded dedication to back-to-the-land living. Erin tells us that she won a prize, $250 for "Best First Garden by a Beginner," for her raised garden. A photographer is coming the next week from *Harrowsmith* magazine.

"I don't know," she says, a worried look on her face. "There's not much growing yet in the garden for pictures."

Dan and Erin became summer friends. We invited them for dinners, we joined them for kayak rides. We watched each other's families grow up. After a few years, they moved to Ryder Lake, near Sardis, wanting the amenities for their children that a larger community could provide. We continued to see each other every summer. They bought property on Harrison Lake, their ties to our lake replaced by their new retreat. But their love of Chilliwack Lake remained.

In August 1987, we sit around the table with Dan Coulter looking over sketch plans for a log cabin for Patrick and me. We are sharing Cupola Lodge, Gran and Poppy's two-room log cabin, with Chris and his family, taking weekends by turns. We're now ready to build our own place. We have had this dream since we were married and spent our honeymoon here. We have the spot picked out, just to the east of the boathouse, where the land rises steeply from the beach.

We, of course, have to locate our cabin keeping in mind that we only own to the high water mark. I have a file three inches thick of court cases that discuss the ownership of land on waterways. There are differing positions across Canada, depending whether the waterway is a lake, river or ocean. And also what province you live in and what provincial laws are in place. In British Columbia, it isn't black and white. The general rule is that a landowner

Our neighbours, Dan and Erin Coulter, moved to live full-time in the Post Creek Community in 1982. They embraced their pioneer life with no running water, electricity or telephone. Courtesy of Coulter family.

owns to the high water mark. This can create some confusion. Some years, the water comes right up to our cabin steps. Other years, the lake rises very little. Some court decisions rely on the vegetation mark as well as the high water mark. A landowner can claim ownership at the point on the beach that vegetation grows. Kevin, who followed his mother's footsteps and became a lawyer, has also considered this question. He and I agree that we should stick with the high water mark. This is the view commonly held by the public and easy to explain.

In locating our cabin, we have chosen a spot well up from the beach. Our cabin will sit up high taking in the view down the lake. Dan shares our vision and will build our log house. We pore over the plans, trying to think through the details. A mudroom for the wet rain boots, a central island in the kitchen for communal cooking, a wrap-around porch. Our boys are still teenagers but we are planning extra bedrooms for their families, a bunk room for grandkids.

Dan tells us his ideas.

"How about if I carve a tree trunk to make the bottom steps of the stairway?"

"I can carve the mantle for the fireplace from a big slab of cedar leaving the bark on. It will look real authentic."

We love his enthusiasm.

Patrick and I and the boys spend two summers clearing the land. Although the trees are not more than thirty centimetres in diameter, the forest is thick. Once we have the trees cut down, we chop off the branches. We drag the branches into piles for burning. The job seems never-ending. We work every weekend, all day, with a couple of hours off for water skiing. The boys' friends are put to work. Our friends are put to work. Once we finish the clearing, Dan builds and pours the foundation.

Patrick and I bring our gin and tonics and two folding chairs and sit on our foundation looking out at the lake. We look at each other and smile.

"Worth all the hard work, right?" he says.

"It sure is."

We visit Dan's work yard in Chilliwack every weekend starting in March the following year. Dan sourced Douglas fir logs in Squamish and brought them to his yard last fall. He wanted the logs to have time to cure. Now, he has men peeling the logs, getting them ready to stack in place. Our log cabin is going up log by log in his yard. Dan wields his chainsaw like a sculptor. He cuts the doorways, archways, window frames with the eye of an artist. One smooth arc, never needing correction. By the end of May, the cabin is finished, sitting ready in Dan's yard to be numbered, disassembled and piled onto a flatbed truck.

On June 15, 1989, the trailer of logs arrives, along with a crane, and Dan and his crew. The truck has to manoeuvre the last eight gravel kilometres on the Chilliwack Lake Road in low gear. Patrick and I are waiting at the entry to our access road. We are anxious to see whether the truck can manoeuvre down our narrow road. Patrick has trimmed back the trees along the sides to widen the road. We hold our breath as the truck pulls the trailer around the bend and through our second gate. A couple of scrapes but the truck gets through.

In only two days, our cabin is reassembled. We are ecstatic! Our dream is taking shape.

Now the finishing work begins. Our contractor, Ralph Belisle, also lives in Post Creek. Kevin and Michael, and their friend, Jon Mawer, spend the summer as carpenter's helpers under Ralph's guidance. They live in my grandparents' original cabin, Cupola Lodge. Patrick and I arrive every weekend and work as the cleanup crew. We gather up scraps of wood and pieces of metal. I am constantly bent over picking up nails and screws. I am paranoid that someone might end up with a nail in her foot.

On August 1, we cut the ribbon on the back door of our new cabin. We are dressed in our best festive attire, straw boaters and baseball caps, carrying an array of glasses and a bottle of champagne. We parade through the door, blowing horns and whistles. We christen the cabin "Creekside," given that it is the closest to Cupola Creek. We crack open the bottle of champagne and we toast what we've accomplished together—Dan, Ralph, Jon, Kevin, Michael, Patrick and I. A lifelong dream is fulfilled: our own cabin at the lake.

Our extended family now has three cabins. Our closest neighbours are blood relatives. Like all neighbours, we don't always see eye to eye. Conflicts inevitably arise. We have two ski boats. We share one dock. We come to the solution of parking our boats on opposite sides of the dock. We remind the grandkids to keep their life jackets and water skis out of the way.

"Let's not leave a trail of debris behind us," Patrick says. "We don't want to be monopolizing the dock."

We work hard to communicate. My nephew, Blake, emails the family. He is having a stag for his best friend the following weekend and hopes we won't mind a bit of noise. We're fine with that, we email back. My brother, Chris, suggests buying a wood splitter. We split the cost as well as the wood. My niece, Hannah, has trouble with their boat. Michael rows over to help her out. For major decisions, like cutting down trees or building sheds or houses, we don't just do what we want, we build consensus. That means giving people time to digest new ideas and having the patience to wait. Every year, we get a little better at compromise.

With the creation of the Chilliwack Lake Provincial Park in 1973, the general public—random, itinerant strangers—becomes our neighbours.

Patrick and I were married on June 23, 1970 and spent three weeks on our honeymoon at the lake. It was idyllic, a string of hot sunny days, the wind warm against our skin. Like the Pied Piper, we lured friends to stay for a few days and say their farewells to us. We had joined the Canadian University Service Overseas and were posted to Zambia to teach in a rural boarding school for two years. We left on August first. Getting ready to leave, I sat on the sand etching the view of the lake into my mind, already feeling its absence.

In the time we were away, the American Fishing Club was sold to the provincial government, creating a park with one hundred camping sites and a boat launch. For the first time, the public had access. We no longer had our slice of heaven to ourselves.

After two years away in Zambia, I am back. I walk down the beach. Dad has told me that the American Fishing Club, Charlie's old place, is a provincial park but I don't know what to expect. As I come closer, I see a cleared swathe down the hill to the beach. Loose rock and gravel litter the hillside and spread out over the sand to the water's edge. This is the boat launch. I take the path leading up to Charlie's cabin. At the top, it hits me: Charlie's cabin is no longer here. None of the cabins from the fishing club are here. In their place are individual camping sites, flattened areas with picnic tables and fire pits, with a separation of a few trees between each site. A track loops its way back into the forest. I walk up this track, stunned to see so many campsites carved into what had been virgin forest. I come to a gatehouse at the entry to the park. A notice of campground rules is posted on the bulletin board.

The pristine forest is gone. Roads, paths, barren clearings, outhouses, signs. I am angry. I want the forest pristine, back the way it was.

We have always had the lake to ourselves. We knew our neighbours: the MacLeods, the Church Camp, the American Fishing Club, the Post Creek Community, the Old Trapper. We knew what to expect from them. The uncertainty of the unknown fills me with unease. My first thoughts are negative. People I don't know will intrude upon my space, my privacy, my peace and quiet.

I force myself to look at the other side of the coin. It isn't easy. But I have to admit: the park will provide access to the general public. People who don't have the privilege of having a cabin can camp here. With a provincial park, the land won't be developed. Maybe the park is a good thing? I'll wait and see. This could all go wrong.

I return to the cabin in late August from shopping in Sardis with Bronwen, Fin and Will. We have done a major grocery shop, taken our clothes to the laundromat where, for an extra two dollars, the operator will move your clothes from washer to dryer, and fold them. Two dollars well spent. We try and find a live rat trap at several feed stores but no luck. When we are just about at the end of our tether, we stop at the Dairy Queen. Blizzards all round. The car is hot and stuffy. We roll down the windows of the car. Murphy is panting in the back.

"Okay, kids," I say. "As soon as we arrive, everyone has to help unload the groceries."

"Can't we go for a swim," moans Will.

"As soon as the car is empty."

We get the job done and I let the kids escape. They throw off their shoes and shirts, dash down the stairs and across the beach and fling themselves into the water. Murphy dashes around with the exuberance of a puppy though he's almost two years old. He races in circles around the kids.

I call the kids in for lunch. They come bounding up the stairs. Suddenly, I notice the quiet. Where is the dog? I go outside and call his name. No response.

"A bunch of teenagers were walking along the beach as we came up for lunch," says Bronwen. Has Murphy followed them back to the campground?

He has never gone to the campground before. I am panicked that I might lose him. Imagine losing your grandchildren's dog! He is so friendly that he would happily go off in a stranger's car.

"You guys stay here, in case he comes back," I say. "I'll go over to the campground."

I push out the canoe and paddle as fast as I can, hoping to see his blond silhouette along the beach. Nothing. I land at the boat launch and start asking questions. Has anyone see a young golden doodle?

"Yes, there was one here. Why don't you check with the park warden?"

I have bare feet and the road is gravel. I wince my way up the hill, my heart pounding. I meet up with a park warden and he radios to the yard. Yes, a blond dog is at the warden's office, now tied up. He'll take me there in his car.

What a relief to see Murphy lying at the warden's feet. The warden gets me a glass of water and we sit and chat. The campground now has one hundred and forty-six sites, a third of which can be booked by reservation through the summer. She is happy that this campground has a reputation for being family-friendly; the rowdy parties go elsewhere. Even though she knows about the private cabins, she encourages park visitors to walk down the beach. I am annoyed that she is encouraging people to stroll down to our beach.

"I wish you wouldn't," I think to myself, but say nothing to the warden. I don't want to antagonize her. At least she's not telling them to set up picnics in front of our cabins.

I feel territorial about our property, not only the land, but the beach and the stretch of water in front of our cabins. I know we only own to the high water mark. The public has the right to the beach below this mark. But if I am sitting on the beach, I resent anyone who invades my space.

Most people appreciate this and keep an appropriate distance. When they encroach, we politely suggest they have their picnic around the corner rather than right in front of our cabin. Sometimes, politeness doesn't succeed and we then have to put up with the intrusion. One of these days, I may have to resort to a "John Hindle."

John was a family friend who owned a house on Lake Okanagan in the interior of British Columbia. One day, a family of four landed their boat on the beach in front of John's house. John went down and politely asked them to move to the campground just around the corner. The dad said he was not moving and told his wife to set up their picnic. She brought out their cooler and folding chairs. Their teenage children snickered. John retreated to the house. He stripped off all his clothes and made up a tray of martinis. He came down to the picnicking family.

"If you are going to stay," John said, "perhaps you will join me in a martini."

"Are you crazy!" the dad exclaimed. They packed up and left John smiling and waving goodbye.

I don't have the guts to do this. I have come up with various rouses: "Sorry about the barking, our dogs are quite territorial;" "I swim lengths in the afternoon and would appreciate you moving your boat;" "We have a baby sleeping, could you turn down the music please?"

I think of the neighbours with whom we share the lake. Those we have grown close to, those who remain an enigma, those who are family, those whom we have yet to encounter. The lake is our common denominator, the thread that weaves us together, the pool into which we all melt. We appreciate it in different ways. It answers different questions in each of us. I try to be open-minded, to respect the relationship each of us has. But I have no patience with those who have no appreciation of its beauty. Those who race muscle boats with unmuffled exhausts. Those who light fires along the shore and leave them smoldering. Those who toss beer cans into the lake. Those are the neighbours I could do without.

TRAIL TO ROAD

In the last week of August, Roger and I speed along Highway One in Rani's Red Racer. We've driven into Vancouver for the day to visit my mother. After our visit, Roger and I stop for groceries at the supermarket and odds and ends at the hardware store. At Abbotsford, we pass rows of raspberry canes, fields of corn, cows grazing in sunny pastures. The Cascade Mountains loom ahead of us at the end of the valley. I can visualize Chilliwack Lake sitting six hundred metres above us, nestled into those peaks.

I turn off the highway at the Yarrow exit. We stop at Wisbey's fruit and vegetable stand. I pick up baskets of local strawberries and blueberries, and some peaches from the Okanagan. Roger chooses a dozen ears of corn: "Jubilee," our August favourite. Our next stop is the ice-cream stand. Double scoops of Bubblegum for Roger and Espresso Flake for me. We sit at the picnic table and let Roger's dog, Dennis, meander about on the grass.

Back in the car, we wind through Yarrow, cross the railway tracks and pass a goose farm. Then we turn east onto the Chilliwack Lake Road. Roger keeps up a steady stream of conversation.

"Rani, you know that woman from Post Creek that you were talking to last week? She said there was a trail to a waterfall near Post Creek. We should go find it."

"Great idea, Roger, how about this weekend?"

Roger remembers everything. He keeps track of details. Every so often he makes my heart sing. "I really like our adventures, Rani," he says.

"Me too." That's an understatement. I wonder if I would have done any of this exploring if it weren't for these grandkids. "I want to explore other trails, too. Like the one the Indigenous people used for hundreds of years along the Chilliwack River."

"There wasn't a road?"

"No, the road wasn't built until fifty-five years ago. I was about twelve when the road went in. Remind me when we get to the cabin to show you the Indian Trail map."

Once we're on the Chilliwack Lake Road, it always feels to me as if we're nearly home. We've left the towns behind, with only mountains ahead of us. Along the river, trucks and campers are parked on the verge. Their owners stand in hip waders in the rippling water, casting for steelhead. Kayakers lift boats off their roof racks and carry them down the bank, veterans mostly, drawn by a fifteen-kilometre stretch of white water, classed II to IV. We pass the Chilliwack River Rafting outfit putting three rafts into the river. A dozen people are standing on the bank, life jackets on, waiting to launch their rafts.

"Rani, we should raft down the river," he says. "That'd be so much fun."

"You know, we *should* do that. We'll get your Dad to organize it."

At the top of the benchland hill, we pass the sign, gate and gatehouse indicating the entry to the Chilliwack Lake Provincial Park. We continue straight ahead onto the gravel forestry road.

"These potholes are getting bigger all the time," I say.

I slow to a crawl and bump along before I turn into our access road, marked by the faded red bar gate.

"Where's the No Trespassing sign?" asks Roger. "It's usually tied onto the gate."

"I guess someone pulled it off."

"Why'd they do that?" says Roger. "Dad and I'll have to put up another one."

Roger jumps out to open the gate and closes it behind me as I drive through.

"I can't wait to dive into the water," Roger says, climbing back in the car.

"Thanks for coming to see Noni with me, Rog," I say.

"I like visiting her even though she never knows my name."

Later that afternoon in the cabin, Roger reminds me about the map. We go to the stairwell where the framed map titled "Indian Territory 1858" hangs on the wall. Hand-drawn by Oliver Wells in 1966, it shows "Village Sites, Historic Sites and Indian Trails." Place names are identified for the Ts'elxwéyeqw, Sumas and Pilalt peoples. The map is decorated with drawings of "Wildlife of Interest to Natives:" fish, eagles, beaver, bear, wolves, deer and mountain goats. My father got a copy from Oliver Wells, of the pioneer Wells family in Sardis, who became an amateur historian and ethnologist focusing on Indigenous history in the Chilliwack Valley. The drawing is meticulous in its attention to detail. The Indigenous traditions are portrayed through the eyes of the 1966 white man, with a veneer of romanticism and racism, too. I know that this map represents a skewed vision, but, for some reason, I still want it hanging on my wall. Nostalgia perhaps?

"See where the trail goes behind the forestry road?" I say. "Do you think we could find it?"

"Sure. Let's try!"

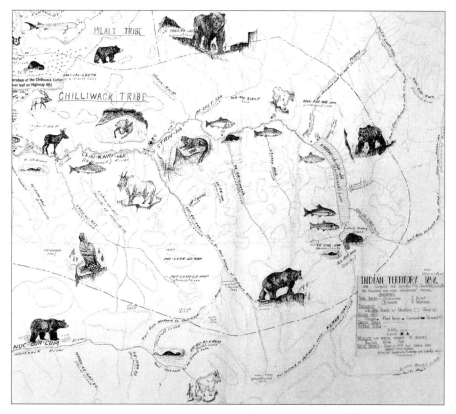

Hand-drawn map by Oliver Wells showing village sites, historic sites and Indigenous trails, circa 1966. Courtesy of the Chilliwack Museum and Archives, AM 362, Map 381.

It's late August and we're nearing the end of our summer. The heat lingers, the air is sultry, the water tepid. Our morning starts with the usual traditions. The "waker-upper"—a morning jump off the end of the dock. The boys are particularly keen. Roger and Fin come out of the boathouse in their bathing suits. They pass by the line-up of adults in folding chairs sipping coffees and reading magazines.

"Bombs away!" shouts Roger.

"Cannonball!" shouts Fin.

This catches Patrick's attention. He leaps to his feet, ready to protect his *Economist*. The boys dive in with barely a splash.

"Hah, gotcha!" says Roger, pointing at Patrick.

Fin gives him a high five. I think the boys have inherited the "teasing gene" from their grandfather.

After breakfast, I suggest that Michael take the kids out in the boat to go cliff-jumping. I want them out of the way. I have put together clues for a treasure hunt, another summer tradition. I don't want prying eyes when I tape the clues in their secret locations. The kids are getting so good at figuring out clues that I make them as difficult as possible and post them at the limits of our property. I use my bike and the canoe to get to the ones furthest away. When I finish, I sit on the dock and wait for the boat to return, enjoying the quiet moment before the tumult of the game.

I watch the speck in the distance become a boat throwing off spray, filled with children. They shout when they see me.

"Hey, Rani. Have you put out the clues?"

They tumble out of the boat.

"We have to stick together," says Nora, the organizer, "so wait for Will."

I give them the first clue and off they go. I sit on the porch, listening to sounds of excitement, argument, frustration and conciliation filtering through the trees. After a while, I see them scrambling into canoes and onto paddleboards, following the clue that will take them around the corner to Otter Cove, where they will find the last clue. "It may be old but can still push logs."

I go to the tool shed beside the peavey to wait, the treasure in my lap. A box of Purdy's chocolates. I can hear their voices getting louder as they run towards the woodshed. They pounce on me, pleased with themselves, as they tear open the Purdy's.

"Next time, Rani, you need to make the clues harder," Roger mumbles, a salted caramel in his mouth.

"The hardest was 'It connected the lake with the valley,'" Fiona says.

"We thought of our access road," says Bronwen. "But we couldn't find any clue on the road."

"Then we remembered the path you call the Indian Trail that goes to the campground," says Nora.

"Man, what a long time it took to find the clue!" says Fin.

"You know, though, that what I call the Indian Trail isn't the first trail. I still want to find the original one," I say.

"Yeah," says Roger, making the most of his insider information. "Rani and I were looking at the old map on the wall at the bottom of the stairs. You can see where the trail used to be."

"Have any of you heard about the residential schools in Canada?" I say. "It's been in the news a lot lately. Something called the Truth and Reconciliation Commission." I have been thinking that I'd like to have a conversation with the kids about this report now that I have finished reading it.

"I think so," says Bronwen. "What's it about?"

"Well, Indigenous children were forced to go to schools. They couldn't stay home with their parents. They had to speak only English. They were punished if they used their own language," I say.

"That sounds awful," says Fiona.

"Yeah, I don't like that," says Nora.

"I know. But it's part of our history that we need to know more about," I say. "There was a school in Sardis. At Coqualeetza. It's an archives now and where I went to do my research on Indigenous history."

Fin looks at me thoughtfully like he's trying to process this information.

It's nearly the end of August. The Indian Trail adventure is set for next weekend before the kids head back to school. While the grandchildren savour their last days of summer freedom, I head down to the Chilliwack Archives to explore the history of the trails to the lake. By now, I greet Shannon the archivist like an old friend. She introduces me to her son, who is working as an intern at the archives for the month of August.

Flora Leisenring has given me the name of a book that describes the early routes through the Cascade Mountains. *Nooksack Tales and Trails.* I ask Shannon whether she has a copy. She brings it out along with *The Pathfinder: A. C. Anderson's Journeys in the West.*

"Those early transportation routes were all about trade," she says. "Fur or gold."

The Fraser River was the first highway into the heart of the province. The river coursed smoothly for 160 kilometres inland from the Pacific coast to Fort Yale, where the steep canyon walls of the Fraser River valley stopped any further travel into the interior. The Cascade Mountains stretching from northern California into southern British Columbia created a nearly impenetrable wall. Footpaths were the only option for those who wanted to continue inland.

"Shannon, I'd like to get a better sense of the Indigenous trails in the early 1800s."

"Do you have the *Indian Territory of 1858* map drawn by Oliver Wells?"

"Yes."

"What about the one drawn by Chief Sepass?"

"No, I haven't seen that." I didn't even know it existed.

She brings out the Sepass map and the Oliver Wells map, too, so I can compare.

"Let's go back to the *Stó:lō Coast Salish Historical Atlas,*" she says. "It has a map of trails before 1866."

None of the maps are to scale, so it's difficult to pinpoint the exact location of the trail, but I get a general idea. I study the trail at the northern end of the lake. That's the section of the trail that I want to find. It just might still be there.

I have finished the Truth and Reconciliation Commission report. The Calls to Action interspersed within the chapters "Legacy" and "Challenge of Reconciliation" are powerful messages. The recommendations concerning Education for Reconciliation strike a chord. To develop and make available "resources on the place of Aboriginal peoples in Canadian history, and the history and legacy of residential schools." Perhaps here is a place where I could contribute.

The Labour Day long weekend arrives. Time for our final adventure of the summer. On Friday night, I pour over my copies of the Oliver Wells and Chief Sepass maps. I compare them to the new Skagit River topographic map that I ordered online. I lay the three of them out on the dining table. Patrick, Kevin and Michael gather around.

"Where should we start?" I ask.

"I think the topography points to going towards the Post Creek watershed, the closest creek south of the lake," says Kevin. "That's the natural trail location."

"And where the trail to Lindeman Lake is," says Michael. "The old trail should be close."

In the morning, it's raining but the kids' spirits are not dampened. They pull on gumboots, slickers and hats.

"What're we looking for, Rani?" asks Fiona, wanting to know what to expect before she sets out.

"We're first looking for a cleared track, maybe ten feet wide, that used to be the original road into the lake," I say. "Then going off that road, we're looking for a trail that leads up to Post Creek."

"It could be faint. Like a deer path," says Patrick.

We splash through the puddles down our access road, then along the forestry road. We veer off the forestry road into the forest, following the topographic map that Kevin has pulled up on his phone. We push the branches aside and make our way through the woods. About thirty metres in, we come to a cleared track about three metres wide.

"This is it," I say. "The original road."

"Really? You drove on this?" asks Fin.

"It's so narrow," says Will.

"Let's follow it," says Roger, and we forge ahead on the track.

"Hey, is this the trail?" asks Bronwen. We stop to have a look at a faint path snaking its way through the forest.

"No, that's not what we're looking for," says Kevin, looking at his map. "We have to go further towards Post Creek."

"Yeah, I think that path goes to the ridge and over to Flora Lake," says Michael.

We carry on through the woods and come to the trail that goes to Lindeman Lake. Well-trodden, about one metre wide, it follows along the bank of Post Creek, exactly as shown on the map.

"Is this it?" asks Will.

"I don't know," I say.

"You know, Mum, I think the trail to Lindeman Lake is probably in the exact same location as the Indian Trail," says Michael.

"But how do we know it's the same?" says Bronwen, the stickler for certainty.

"We don't know for sure," says Patrick. "But it makes sense that it's in the same place."

"So, hundreds of years ago, Indians were walking on this trail?" says Fiona. "Cool."

At the end of the weekend, I return to the Chilliwack Archives for further research on the road into the lake. By closely comparing maps from various times, I conclude that the Indian Trail was most likely the basis for the roads that came after. From Vedder Crossing, the trail once linked the Indigenous peoples' villages along the Chilliwack River with Chilliwack Lake. From there, the trail continued east following creek valleys through the Cascade Mountains to Fort Hope, a Hudson's Bay Company trading post, now the city of Hope. Indigenous trappers travelled the trail to Fort Hope to exchange fur pelts for food and supplies.

Then the miners came.

"Gold Discovered in the Fraser Canyon" blared the newspaper headlines. It's the summer of 1857. News spreads so fast that within the space of a few months, gold-diggers from California are converging on the lower Fraser River in droves. The miners press for access to the upper reaches of the Fraser River. In the United States, thirty-two kilometres south of the border, the town of Whatcom in Bellingham Bay seizes on the chance to be the stepping-off point for miners heading north. County leaders raise money to build the Whatcom Trail to connect Whatcom with Chilliwack Lake via the banks of the Nooksack River.

Online I find a report, *Whatcom Trail of 1858*, prepared by the Skagit Environmental Endowment Commission to commemorate British Columbia's one hundred and fiftieth anniversary. It contains excerpts from historical documents: diaries, maps, and aerial photos.

Ten thousand miners are camped at Whatcom, seduced by the promise of riches. Anxiously they await the completion of the Whatcom Trail. A team led by W. W. DeLacy, an American engineer, is hired and trail construction begins. By late June 1858, word comes back that the Whatcom Trail has been completed to Summit Lake (also known as Chilliwack Lake). As reported by DeLacy:

> We arrived here Summit Lake today, all right. The men are in good spirits, although we have had nothing but coffee and flour for ten days. We make a raft today to cross the lake on.

The *Northern Light*, a Whatcom newspaper, announces that a hotel has been built on Summit Lake to accommodate travellers and their horses.

> Tolerable fair meals, for one dollar each can be procured. A ferry is available to cross the lake lengthwise, about a seven-mile journey.

I stop reading and sit back in my chair. I never imagined a hotel at the lake in the mid-1800s, nor a ferry crossing. Ten thousand men? Crossing my lake?

The Whatcom Trail starts at Bellingham Bay, travelling northeast to meet the Chilliwack River at Chilliwack Landing where a ferry service takes men, horses and supplies across the Chilliwack River. The trail then joins the Hudson's Bay Brigade Trail along the north bank of the river to Chilliwack Lake. A ferry carries miners from the north end to Depot Creek at the south end. The trail continues up the Depot Creek valley to meet the Skagit River and on to Hope in the interior of British Columbia.

Reports of completion of the Whatcom Trail are exaggerated. (I wonder if the hotel is pure fabrication.) Speculation spreads that the trail will never be finished. The *Northern Light Extra* of August 19, 1858 announces, "Cheering News! The Trail Through! No Humbug this Time!" Champagne flows and one hundred guns are fired from the hill above Bellingham Bay. It is declared that the trip from Bellingham Bay to Hope with pack animals will take ten to fifteen days.

Although the Whatcom Trail is "finished," it is nearly impassable in places. Within months, the balloon of the gold rush bursts and miners and businesses pack their bags and leave. Dreams of riches evaporate and the streets of Whatcom become as deserted as the paths of the Whatcom Trail. The hotel and ferry on the lake must have been very short-lived if they existed at all.

It is the Boundary Commission survey teams that revitalize the section of the Whatcom Trail from Vedder Crossing to Chilliwack Lake. The American survey team arrives first, in the fall of 1858. Further upgrading of the trail is completed in the spring of 1859 in time for the arrival of the British survey team. For five years, the trail provides the access for the international survey teams through the Cascade Mountains.

At the turn of the twentieth century, gold is discovered in the Cascade Mountains. I turn to my most constant reference companion, *In the Arms of the Mountains: A History of the Chilliwack River Valley.* I originally found the book at the archives, but I have since bought my own copy so that it's always at my fingertips. The Boundary Red Mountain Mine is established on the American side of the border. Supplies required for the mine are most easily delivered from Vedder Crossing on the Canadian side of the border.

In 1901, the trail from Vedder Crossing up the Chilliwack River sixteen kilometres to Slesse Creek is upgraded and renamed the Mount Baker Road. Wagons can now travel the Mount Baker Road to Slesse Creek. From there, miners and pack trains ford the Chilliwack River and climb a narrow track up the mountain eight kilometres to the mine. Chilliwack packers carry supplies in and gold bricks out from the mine over the Mount Baker Road by wagon and horseback. Settlers developing homesteads along the Chilliwack River valley also make use of the Mount Baker Road. East of Slesse Creek, the Hudson's Bay Brigade Trail remains the only transportation route to Chilliwack Lake.

In the mid-1920s, the Mount Baker Road takes my grandparents by car as far as Bell's farm. The Brigade Trail takes them by horseback the remaining thirty-two kilometres to the lake. Gran, Poppy, Frances and Rowly leave Chilliwack before dawn for the Bell's farm. They are met with a bustle of activity. Horse wranglers saddle the packhorses. Boxes of supplies are tied as evenly as possible on either side of the horses with thick ropes cinched tight. Some of the packhorses are tied in a line, front to end, to keep them following the pace of the lead horse.

Poppy checks the packs, making sure everything is secure. He is particular about the way the ropes are tied and a stickler for leaving on time.

"Come on, Ed. Time to get moving," he says, as the sun breaks over the horizon.

Ed Bell helps Rowly get settled on his horse. Gran sits patiently on her placid mare, awaiting the long day ahead.

Mount Baker Road was widened to accommodate wagons, circa 1901. In the 1920s, a gravel road extended sixteen kilometres up the Chilliwack River valley. The road was completed through to Chilliwack Lake in 1961. Courtesy of the Chilliwack Museum and Archives, AM 0773.

For the first time, Frances, who is seven, is on her own pony. She is eager to start.

Poppy smiles at her, "Frances, you lead the way."

As the day progresses, the sun becomes a throbbing presence. They stop to rest the horses, let them feed on the grass and drink water in the creek. Frances and Rowly dip their bandanas in the stream and wet the back of their necks. Gran passes them a tin cup and they take turns drinking water from the creek. Clear and cold. They travel across a narrow wooden slat bridge strung across the river with thick ropes. The bridge sways under the horse's weight. Ed leads the way when they ford a creek, assessing which route will give the horses a solid footing. The riders' legs feel the bite of the icy water.

They come to the last hill before the benchland. The horses are strung out along the switchbacks going uphill. Old Jack, the horse ahead of Gran, is carrying the boxes of eggs and bags of vegetables. Suddenly, Old Jack careens off the trail through the scrub pine, surrounded by a swarm of wasps. Coming up next, Gran's horse starts shaking its head, nervously side-stepping up and down the trail. She calls out to Ed. The men leap into action. Some head their horses back down the trail to intercept Old Jack. Others dismount and retrieve the boxes of supplies now strewn down the hillside. Ahead, Frances and Rowly turn around to see what is causing the commotion.

Poppy calls to them, "You two wait at the top of the hill."

The men gather the horse and the supplies. They manage to calm Old Jack and tie his packages more securely on his back. The troop continues on, arriving an hour later at Charlie's cabin. When they open the box at the end of the trip, they find only a few eggs broken.

I know that as a teenager, Mother rode on a logging railway, with only a few kilometres on horseback at the end of the trip. I want to find out when this railway was put in. And why it stopped operation. *In the Arms of the Mountains* provides some leads to follow up in the archives of the *Chilliwack Progress* newspaper.

In the late 1920s, the Campbell River Logging Company acquires logging rights in the valley, constructs a rail bridge across the Chilliwack River at Vedder Crossing and lays track along the south side of the river. As the old growth Douglas fir is logged, the rail line is pushed further and further east up the valley. The rail line continues eastwards to within four miles of Chilliwack Lake.

From the early 1930s, Gran and Poppy, Rowly and Frances travel by logging railway and packhorse trail. The fire of 1938 destroys the railway as well as logging in the valley. The camps are dismantled and the rail lines pulled up. This is the end of overland travel to the lake for twenty-two years. The family now comes in by float plane. During the Second World War, Gran and Poppy fly in with the MacLeods. Mother is with the Women's Royal Canadian Naval Service stationed in Halifax. Dad is with the Canadian Artillery stationed somewhere in Europe. Dad's first trip into the lake isn't until 1948, the year I am born.

In 1956, the Forest Service becomes interested in completing a road to Chilliwack Lake to re-establish access to logging in the Chilliwack River valley. Later, in 1964, my mother would take on the Forest Service in the extension of this road. Now, in the spring of 1957, survey crews marking the road right of way make their way to the north end of the lake. Walking in the woods one July afternoon, Poppy sees red marking tape on trees at the shoreline only a few hundred feet from the cabin. Poppy follows the markers back onto the benchland and finds a camp of surveyors.

"Who are you?" Poppy asks. "And what are you doing?"

"We are with the Forests Branch," one replies. "We are surveying the road."

"You can't put the road here. This is private land, not Crown land," Poppy says.

"Well, that's not our decision," the lead engineer replies. "We just put tape markers where our maps show the road is supposed to go."

"Well, we'll see about that," says Poppy.

Men at Bowdenville Logging Camp in the Chilliwack Valley, circa 1938. The steam loco-motive hauled logs along the logging railway. Courtesy of the Chilliwack Museum and Archives, PP502576.

I see from Poppy's correspondence files that as soon as he is back in Van-couver, he phones the Forests Branch. He is not happy with the non-com-mittal response. He writes a letter to the deputy minister of the Forests Branch to register his concern. The road should not be on private land, his land. The road should use the Crown right of way at the north end between the American Fishing Club and the MacLeod property. After a whirl of com-munication, the Forest Service finally agrees that the road will be located on the Crown right of way and not on Cupola Estates land.

I follow the pace of the road construction in the *Chilliwack Progress* newspaper archives. Advancement up the Chilliwack Valley continues as money and labour are available. In early 1959, the paper reports that activity has halted due to lack of funding. However, later that year, the Forest Ser-vice crews, assisted by inmates from the prison camps, complete the road to within sixteen kilometres of Chilliwack Lake.

Finally, in 1961, the road is finished through to the north end of Chil-liwack Lake. The *Chilliwack Progress* reports that the "Mount Baker Trail, once a lively access route for gold miners prospecting in the border country, is getting its name changed." Chilliwack Lake Road starts at Vedder Crossing and continues thirty miles to the northern end of Chilliwack Lake. Initially gravel, "The highways department plans to pave the first mile this season."

Poppy has qualms about the road. The hidden jewel of Chilliwack Lake will be revealed to whoever wants to make the drive. An asphalt ribbon will wind its way through the forest. Broken trees and rubble will be left in the wake of the bulldozers.

John MacLeod remembers hearing large machines at the top of the benchland during the summer of 1961. When he goes to investigate, he sees Poppy trudging through the forest waving his arms and calling out to the bulldozer operators. The machines are heading in the wrong direction, he yells. He asks them to look at their survey plans. They get out the plans. They are indeed off course. They turn the machines and take the correct trajectory to follow the forestry right of way.

When I am fourteen and Chris is twelve, our family spends the month of July clearing a mile of dense forest to create our access road from the newly completed Chilliwack Lake Road through our property to Cupola Lodge. Dad uses a chainsaw to cut down the small fir trees and buck them up for firewood. Chris and I gather the branches and toss them into the forest. We pile the firewood rounds into wheelbarrows and push them back to the woodshed. We dig up the small stumps with pickaxes and shovels. The large stumps require more muscle. Dad winds a chain around the largest and uses the van to drag them out of the ground. The routine is to work in the mornings until lunch. Our afternoons are free.

It is hot with mosquitoes and black flies buzzing around my head. My skin is sticky with sweat and a fine layer of sawdust sticks to my skin and gets under my shirt. The dust itches and I get a rash on my back. My arms and legs are covered in scratches. Tiny slivers sit just below the surface of my skin. Gran brings lemonade to us mid-morning. Tart, with honey added for a hint of sweetness. Dad tells us it's time to knock off. Chris and I sprint down the track, racing to see who will be the first one in the water.

Our first drive on the newly completed Chilliwack Lake Road is in July 1962. We leave Vancouver at six in the morning. Our supplies are packed into the second-hand one-ton Chevy panel van, pea green with rusty bumpers. Mother and Dad sit in the bucket seats in front; Chris and I on the wooden benches bolted to the floor along each side of the van. The trip starts off smoothly enough, following the highway from Vancouver via Langley to Chilliwack.

Once we turn off at Vedder Crossing onto the Chilliwack Lake Road, the route becomes a weave of twists and turns. A few kilometres of pavement and the road turns to gravel. There are no windows in the back of the van. Even with the front windows rolled down, it is stifling. Dust sifts up through the floorboards, coiling in circles through the back of the van driven by the breeze from the windows. Everything is soon covered in dust. The smell of gasoline mixed with fertilizer and a panting dachshund combine to make me nauseous. Dad stops every few hours so we can stretch our legs and I can breathe fresh air. We pass three of the prison camps and see young men in blue uniforms walking along the side of the road, shovels and pickaxes in their hands. The road goes right through the last camp, Centre Creek.

The family tumbles out. Chris and I hang around the van. I am fourteen and feeling especially awkward. Inmates walk by and stare at us. Mother and Dad go into the camp office to register our car. I am relieved to get back into the van and continue on our way. The road narrows, now barely a single lane, with boulders and potholes. The van crawls along at a snail's pace.

We come to a creek. The road goes straight through. There is no bridge. Dad stops the van, gets out and considers what route he should take. He presses on the accelerator and the wheels spin as we bounce across the creek bed. I hold on to the edge of the bench to keep my head from banging against the side of the van. We make it across.

Dad laughs, "That wasn't so bad."

Another hour and we come to the final crossing, the Chilliwack River.

"This is the last crossing before the lake," he says.

"Hang on."

He takes a line and floors it. Halfway across, the van stalls. Water is leaking in through the doors. Panic blooms in my stomach. What if we can't get across? We are in the middle of nowhere. We haven't seen another person for hours.

"Okay, everyone out and push."

We step out of the van into the river. The water is cold and moving fast. I feel my feet being pulled out from under me. Chris, Dad and I go to the back of the truck and push like demons when Dad yells "Push." Mother is at the wheel. The van starts up and bounces over the boulders. Mother guns it to the far side of the river. We all struggle not to fall flat on our faces.

"We made it!" Chris shouts.

Dad lets out a sigh of relief and walks around the van, checking for punctures in the gas tank. All is well. The road continues in switchbacks up the steep grade in front of us. Dad has the van in low gear as it grinds uphill, spitting out gravel in the deep holes. At the top of the hill, Chris and I start to smile. We can see the mountains down the lake. We can see the Cupola. We are nearly there.

At about five in the afternoon, after a ten-hour slog, we arrive at the cabin drowsy, dusty, and depleted. We strip off our clothes and plunge into the water. The cold seeps into my pores. I duck my head under and open my eyes to the blur of pale green. The silky water envelops me.

As a teenager, I don't grasp the significance of the changes that the road will bring. The trip from Vedder Crossing is a tortuous journey. Not for the faint of heart. But over time, it will get easier and take less time.

Once our access road is serviceable, Mother and Dad finalize their plans to build their summer cabin. They choose Pan-Abode, a pre-fabricated squared-log construction.

In early July 1963, a truck pulling a flat-decked trailer stacked with numbered logs trundles down our access road. By the end of the summer, a Pan-Abode log house attached by a porch to a one-room sleeping cabin is completed. Mother insists on a stone fireplace and indoor plumbing. She spends days alongside the Chilliwack River searching for the perfect stones for the fireplace. One by one, she brings them back in the van. The indoor bathroom has a toilet, sink and shower. A Paloma gas water heater delivers instant hot water for the shower and tap in the kitchen.

I am sad when we first move to the Sandpiper. I no longer wake up to Gran in the kitchen and Poppy shaving on the front porch. I imagine my grandparents, alone in the Cupola, missing me too.

Over time, the road is improved, bridges are built across the rivers, sections are straightened and transformed in fits and starts from gravel to pavement. In 2003, the road is fully paved to the Chilliwack Lake Provincial Park Campground. In addition to the one hundred and forty-six campsites in the campground, there is a steady influx of day visitors. On busy weekends, picnickers are crowded cheek by jowl on the beach in front of the campground. The noise of speedboats and water scooters reverberates across the water. Sound systems from the boats and the beach blare music we can hear inside our cabins. The trails to Lindeman and Flora Lakes are crowded. We find litter on the sides of the forestry road and on the shore.

The road is a double-edged sword. It brings people and progress, both of which puncture our peace. On the other hand, we can now drive to the lake whenever we want. We are not limited to a once-a-year visit. We are not held hostage by clouds, waiting in Chilliwack for an opening in the skies that will allow us to fly in. We are not restricted to the supplies that can be brought on

A wooden span bridge across the Chilliwack River wide enough to accommodate a horse and buggy, circa 1895. The Vedder Bridge linked Sardis and Vedder Crossing with Cultus Lake. Courtesy of the Chilliwack Museum and Archives, P3504.

a six-seater float plane. My parents can truck in prefabricated logs to build their Pan-Abode cabin.

These trails and tracks and roads have been agents of profound change. They have spurred our move from simplicity to increasing complexity. Sometimes I wish I could go back to the soft tread of footsteps on the Indian Trail. I envy the roadless wilderness that Poppy knew. I can no longer fully imagine such a world.

But would I be happy without the ability to do this research that has meant so much to me over the course of this summer? Without my family able to drive here in a matter of hours?

The truth is, we are trapped in the world we are born into. Mine is crisscrossed with roads and packed solid with people—and not always people I choose.

Prison Camps

It's 1963, the beginning of August, and we're on our way to the lake for a month. I'm fifteen and Chris is thirteen. The Chilliwack Lake Road is just barely finished, still full of potholes and rough gravel. Our panel van labours along, never reaching a speed of more than twenty miles per hour. Three-quarters of the way to the cabin, we take the turn-off to Mount Thurston Correctional Centre, one of five "forestry camps" built in the Chilliwack River valley in the 1950s and 1960s. Dad has heard that Mount Thurston Correctional Centre has a sawmill and planer mill. Being in the lumber business all his life, Dad is curious. He asked around and found out that prisoners at Mount Thurston produced the lumber used to build the prison camp buildings, the decking for bridges along the Chilliwack Lake Road—and picnic tables. A picnic table made at a prison camp? He decides we have to have one of those.

"Might as well pick up that picnic table now, on the way," says Dad.

"Really Gary, do we have to?" says Mother.

"Ah Dad, let's just get to the lake!" moans Chris.

Dad stops the van in front of the gate: the gate and the sign "Mount Thurston Correctional Centre" are the only evidence that we are entering a prison camp. Dad tells the guard why we are here. The guard waves us through and instructs us to ask at the administration office, a wood-clad building sitting at the edge of a semi-circle of similar buildings. The prisoners sleep in ten or twelve basic one-room structures, the white paint faded and peeling, each with three steps leading to a covered porch at the entry and a sloped shingled roof. A flagpole is planted beside the office, the Canadian flag fluttering listlessly. At the far side of the semi-circle, a well-trodden path in the sparse grass leads to the dining hall.

At fifteen, I am acutely aware of the danger lurking in the shadows of the camp buildings. Men stare at the van as Dad swings around and parks in front of the office. Some look not much older than me. My mind races over the crimes they might have committed. Stealing cars, robbing banks, rape, murder. Not those kinds of crimes, lesser crimes, my mother says, but even so, these are bad boys.

"Can I buy a picnic table?" Dad leans out of the van window and calls out to a guard going up the steps of the administration office. Mother and Dad get out of the van and approach the guard. Although my father is six feet tall with an athletic build, his manner is amiable and relaxed. The guard on the other hand, short and squat, is belligerent.

"Whad'ya want it for?"

"For our cabin up at Chilliwack Lake," Dad says mildly.

"Well, I don't know if we have any."

Mother's brows tighten and her face flushes.

"Maybe you could check?" Dad says.

The guard barks at a passing inmate, "Hey, Freddie. Go to the storage shed and see if you can find a picnic table. This guy wants to buy one."

Chris and I are standing by the car, scuffing our runners in the dirt, sensing the tension and wishing we could get going.

Two prisoners come back carrying a box marked "Picnic Table, Cedar."

"Well, whad'ya know?" the guard says. "It'll be ten bucks."

Dad hands over the money. The inmates help him load the box into the van.

"Who knows whether all the pieces are there," Mother mutters as she gets back into the vehicle.

It's early September and I am on my own, everyone else is back to work and school. A breeze shivers off the water and the leaves on the cottonwoods flutter, yellowing and curling along the edges. September, and the lake is quiet, with only occasional echoes of summer exuberance. The sun is hiding, the curtain of clouds occasionally parting for a cameo appearance. A shaft of light captures the cottonwoods along the shore, their leaves starting their annual rotation to gold. The only sounds are the scurry of a chipmunk across the deck and the whisper of wind through the leaves.

I am returning from a shopping trip to Sardis stocking up in anticipation of the boys' families this weekend, as well as Patrick. Kevin and Michael block off a few fall weekends to bring their families to the cabin. The rest are taken up with soccer and field hockey games, and sleep-overs.

This summer, I have passed the Ford Mountain Correctional Centre dozens of times—back and forth into Sardis, where I go to buy groceries, do banking, pick up flashlight batteries, stop for gin and wine at the liquor store, all the mundane chores that keep our summer life running smoothly. Ford Mountain is the one remaining prison camp on the Chilliwack Lake Road. I pass it too on my trips to visit the Coqualeetza Centre and the Chilliwack Archives where I have devoted so much time to learning about the lake's history.

Back and forth on the road. After a while it becomes rote. I sometimes get to the end of the Chilliwack Lake Road where I turn right into Sardis and I can't remember a thing about the past forty kilometres. My Prius must have driven all on its own.

Sometimes, however, changes in the valley snag my attention. The diner near Vedder Crossing is suddenly yellow. I catch a glimpse of a "wood for sale" placard along the road. There's a new sign at the fish hatchery advertising opening hours.

Whoa! I can see the prison fence! The forest between the road and Ford Mountain Correctional Centre has been cut down. When did that happen?

These "forestry camps" have been here in one form or another for fifty years, but suddenly, now, Ford Mountain is "in my face" and just ten kilometres down the road from our cabin.

When were the first prison camps built along this road?

Why were they placed here, in the wilderness?

In British Columbia, the Ministry of Public Safety and the Solicitor General administer correctional centres. On the ministry website, I click on *History*. The first prison in British Columbia was Oakalla Prison Farm, built in 1912. Prison policy initially focused on containment. But in the late 1940s, the approach shifted towards rehabilitation. Oakalla Prison Farm bought property on the Chilliwack River to build a rehabilitation centre. No reason was given as to why the Chilliwack River valley was chosen for this initiative. "Forestry Camps," as minimum security prisons were called in the early 1950s, were established to teach trades to inmates to facilitate their eventual return to society. The Tamihi Correctional Centre, built in 1956, was the first of five prisons built in the Chilliwack River valley. Mount Thurston, Centre Creek, Pierce Creek and Ford Mountain followed in quick succession. These prisons were intended as transitional facilities for men nearing their release date with minimal security: no barbed wire around the prison, no locks on the doors and no guns for the guards.

I read that during the late 1950s, the road from Vedder Crossing to the lake was constructed by the prison camp inmates. Until the road was finished in 1961 and we drove in ourselves, we were oblivious to the presence of the prison camps. Up till then, we had arrived by float plane: we didn't see the prison camps or even know they were there. The closest was eight kilometres from our cabin.

Once the road was finished, we drove past the prison camps. We saw road gangs by the side of the road. The young men wore blue shirts and pants and carried shovels and hoes. Dad slowed down so he wouldn't create a cloud of dust. Mother looked straight ahead. Chris and I were in the back of the van straining to catch a glimpse of the prisoners. I didn't even have time to wave, they were gone in a flash. As a twelve-year old, I couldn't comprehend what life in a prison or the prisoners themselves would be like. I was protected inside a cocoon of privilege.

I hear a "Hello, anyone home?" from behind the cabin. At fourteen, I have finally rejected my mother's bowl-shaped haircut and am letting my blond hair grow fashionably long. I unwind myself from the hammock stretched between the birch trees beside the cabin and put my book down on the grass.

Around the corner strides a correctional officer wearing a brown uniform with BCGS (British Columbia Gaol Service) in a patch on his sleeve and on the brimmed cap on his head.

"Someone escaped from the Mount Thurston camp," he says. "We've set up blocks on the Chilliwack Lake Road near Vedder Crossing."

He tells us that two men walked away from a work party the day before, wearing blue denim pants, blue shirts and black work boots. One of the men has two birds tattooed on his right forearm. The officer stays for a glass of lemonade.

"Nothing to be concerned about," he says as he leaves. "Escapees usually head down the road to Sardis, not up towards Chilliwack Lake. But let us know if you see anything suspicious."

The MacLeods come by later that afternoon. They've had a visit from the officer, too. Flora and Mother are worried.

"What would we do if we saw these men?" asks Flora. "We don't have any way to contact Mount Thurston. We'd have to drive to the prison camp."

"We're at the end of the road," Dad says, thinking this is a comfort. "There's no way out. They won't come here." I listen to what Dad says. He's calm and rational, a foil to my mother's volatility. He knows things. I trust him to keep me safe.

The next day we hear that the two prisoners were captured and returned to Oakalla Prison Farm in Burnaby. Even so, news of this escape unsettles me. I check that the doors are locked when I go to bed. I startle at unfamiliar noises in the night. I keep my worries to myself, arguing silently that a prisoner would never come to our cabin. Dad said so; it must be true.

I put on a brave face. I put live bait on fishing hooks. I chuck drowned mice into the wood stove. I slit the stomachs of the fish I catch. I don't like doing any of these things, but I do them. I also keep a baseball bat under my bed.

A screech of splintering wood and a crash. I waken with a start. We finished building our cabin a couple of years ago and we're here in the fall with Michael, who is in his last year of high school.

"Patrick," I jostle his shoulder. "Someone's downstairs."

He is instantly awake.

"Stay here," he says, grabbing the flashlight beside the bed.

We can hear voices from below. "I found some jackets. Check for boots."

Michael is in the next room, with our yellow Labrador retriever, Monty, sleeping at the foot of his bed. Patrick shouts his name, and Michael staggers out holding a baseball bat. He is eighteen, stocky and strong. Monty bounds for the stairs, barking.

Patrick and Michael stumble down the stairs shouting, "Hey! What're you doing?"

I follow them, keeping my distance.

Before Patrick and Michael can get to the back door, we hear clattering and heavy footfalls down the porch steps. Monty rushes off into the dark, barking furiously. Patrick and Michael shine their flashlights into the night, reflecting ribbons of rain. They catch a glimpse of two figures running into the woods.

The doorframe is twisted and the door is hanging off its hinges. Coats are strewn across the floor. Drawers are upended. We huddle together at the door.

"What was that?" Michael asks, calling Monty back.

"Who knows?" Patrick says, bending to pick up a stray boot and a rain slicker. "Maybe guys from the prison camp?"

Patrick and Michael jostle the door closed and nail the frame temporarily back in place. We're all shaken.

"Does anyone feel like some tea?" I ask.

"No thanks, Mom."

We sit at the kitchen table and talk for a while.

"What should we do?" Michael asks.

"This is why we should have a phone," says Patrick. "We'll have to wait for morning to go to the prison camp."

"We can't do anything more tonight," I say. "Let's go back to bed." We reluctantly head up the stairs, saying good night to each other. I sleep uneasily, startling at the slightest noise. Nobody sleeps very well.

The next morning, Patrick and I drive the eight kilometres to the Centre Creek prison camp. We drive through the prison camp gate and pull up to the office.

"Hi, I'm Patrick O'Callaghan and this is my wife, Shelley. We have a cabin up at Chilliwack Lake," Patrick says to the guard behind the desk. "Last night, around midnight, a couple of guys broke down our back door. Could they be from your camp?"

"Yeah, we had two fourteen-year-olds escape yesterday afternoon," the guard says. "Did you get a look at them?"

"No, it was dark and raining hard."

"They were probably looking for dry clothes and a place to hide out for the night," says the guard. "Thanks, I'll report this and we'll let you know. We've already set up roadblocks on the main road. They won't get far. They never do."

As we get into the car, I turn to Patrick, "Fourteen, that's awfully young."

"Hard to imagine, four years younger than Michael."

This is the first break-in we've had at our new cabin, Creekside. We've had break-ins before at the other cabins, but never when we were in the cabin. When we get back, we tell Michael what we learned.

"I am keeping this baseball bat under my bed from now on," says Michael.

"I feel safer having Monty here," I say.

"Monty wasn't much help. He didn't hear anything until Dad woke us up," Michael points out.

"Still, I like having the dog."

"Those kids were more probably more frightened than we were," Patrick says. In the light of day, it seems like a dream. A bad dream. Michael and Patrick focus on fixing the door. Did it really happen?

Two days later, an officer appears at the cabin to tell us the prisoners were captured the next day, hitchhiking on the Chilliwack Lake Road.

My dad was distraught to hear about our break-in. Bad enough when it happened to him, but another thing entirely when it happened to his daughter and her family.

The following year, the government proposed building a new correctional centre to replace Ford Mountain Camp, about ten kilometres away from our cabin. This upgrade, at a cost of seventeen and a half million dollars, would add one hundred and ten beds, an increase of nearly one hundred per cent. Dad spearheaded a campaign to oppose the new facility. To him, expanding the prisons in a developing recreational area such as the Chilliwack River valley was an outrage. The *Vancouver Sun* printed Dad's letter:

> A Citizen's Group has recently campaigned against the Attorney-General of BC's plan to build a prison which will double the

prison capacity in the Chilliwack River valley. This area now suffers from one escape per week, from unfenced prisons holding murderers, rapists and other violent offenders.

This issue began as a Chilliwack River valley problem. It is now a small battle in an urgent overall problem of concern to all British Columbia citizens. It is up to us, the public, to change the system.

All through July and August that year, my father walks along the beach every day. A fringe of white hair sticking out sparsely beneath a worn tennis hat, blue eyes behind rimless glasses, faded blue jeans rumpled and baggy at the knees. He carries a clipboard with his petition in his hand and a pocket full of pens. He is stubborn and relentless. He sets off walking down the beach towards the campground right after lunch and isn't back until five. He speaks to every person he sees on the beach and visits every campsite. He says people are friendly, they like to talk. He goes up the road to the park entrance. He stops every car that passes.

Dad is following in Poppy's footsteps in his crusade to protect the lake from harm. He wasn't born to it, but he has come to love the lake. This place demands such passion.

His question is simple, "Do you want more prisons in the Chilliwack River valley?" He obtains three thousand, three hundred signatures. Ninety-three per cent are opposed to the proposal. Throughout the fall, he writes letters to the editors of the *Chilliwack Progress* and *Vancouver Sun*, to the mayor of Chilliwack and members of the legislative assembly of British Columbia.

In British Columbia, agricultural land is protected and designated part of the Agricultural Land Reserve. The Agricultural Land Commission must approve the removal of land from the Agricultural Land Reserve. In January 1995, the Agricultural Land Commission holds a hearing at which the attorney general requests the release of land from the Agricultural Land Reserve for the expansion of the Ford Mountain Correctional Centre. Dad attends and makes a presentation. The Agricultural Land Commission denies the request. I hear the news and phone to congratulate Dad.

"It's not time to celebrate yet," he says.

The Agricultural Land Commission decision provides a loophole: "There appear to be other viable sites outside of the Agricultural Land Reserve for the relocation of the proposed prison. It is suggested that these, and possibly other viable options outside of the Reserve, be explored in greater detail."

This galvanizes Dad into writing another flurry of letters to the ombudsman of British Columbia, and to the newspapers. His rallying cry is,

"The public is strongly opposed to a prison anywhere in the Chilliwack River valley."

Dad's concern that the Agricultural Land Commission's decision is not the last word on the prison expansion is well founded. In September 1995, after pressure from the attorney general and the regional district, the Agricultural Land Commission reverses its decision and allows the land to be removed for prison expansion.

I console my father. He fought the good fight and lost. But I am wrong. He is not finished. Dad calls his friend, Thomas Berger, who is a Queen's Counsel and a pre-eminent legal figure in British Columbia, renowned for his representation of the Nisga'a Nation in a case in which the Supreme Court recognized Indigenous rights in Canadian law. Dad asks him to appeal the Agricultural Land Commission's decision. Mr. Berger sends letters questioning the legality of their reversal of the original decision. In January 1996, the Agricultural Land Commission advises the attorney general that their reversed decision is illegal and reinstates the initial decision, disallowing the request for removal of the land from the reserve. The expansion is defeated. Dad, the crusader, is triumphant.

I am of two minds about the prison expansion. I agree with Dad that we don't want an increased prison presence in the Chilliwack Valley. However, prisons need to be put somewhere. If not our backyard, then whose? Maybe this is an appropriate location. Is a natural environment more conducive to rehabilitation than an urban environment? It is always easier to say no to a proposal with potential negative consequences in your neighbourhood than to say yes. This feels like taking the easy way out instead of tackling the problem. Then, in February 2002, the provincial government announces across-the-board cutbacks to government services. Beginning in 2002, ten correctional centres will be closed. Dad learns that prison camps on the Chilliwack River are slated to close. He calls to tell me the news.

"Why are the prisons closing?" I ask Dad.

"Budget cuts," says Dad. "I read about it in the *Vancouver Sun*."

According to a British Columbia 2002 Service Plan for Correctional Centres, the ministry will:

> Consolidate correctional facilities for greater efficiencies and reduced costs. There has been a recent decline in the prison population. Incarcerated offenders have become increasingly more serious and are less suited for open and medium secure centres.

Over the next ten years, all the prison camps in the valley close except for the Ford Mountain Correctional Centre, now run by the Fraser Regional Correctional Centre District.

I read on the ministry website that the Ford Mountain Correction Centre is one of only nine provincial adult custody centres currently in operation in British Columbia. Ford Mountain is a men's multi-level security centre primarily for Indigenous men, offenders with a mental disorder and sexual offenders. It houses up to one hundred and twenty-five men, generally serving less than two years.

We drive by Ford Mountain Centre on every trip to the lake. The security lights, motion detectors, perimeter cameras, double fencing with the outer curved inward are a visual reminder of the prisoners in our midst. It is a far cry from the forestry camps of fifty years ago. The simple cabins surrounded by woods have been replaced by cement block buildings surrounded by razor wire.

I am uncomfortable knowing Ford Mountain Centre is only ten kilometres from our cabin. I am anxious for our safety, especially now that I have children and teenagers to worry about. I try and take comfort in Dad's assurance that escapees will choose the road to the bright city lights and not the one that veers into our dark wilderness.

I have just about arrived at our access road when another sign jumps out at me, "VQ The Creek."

What is this place, I wonder? And why have I never noticed it before?

When I arrive at the cabin, I turn on the computer. In 2010, the Stehiyaq Healing and Wellness Village was opened on the Chilliwack River on the site of the former Centre Creek Correctional Centre. Indigenous leaders worked for years on this initiative, finally receiving the support of the provincial government to open Stehiyaq. Its purpose was to combine the healing practices of Indigenous Elders with modern medicine to assist youth at risk. It was short-lived, but the property was taken up by another initiative.

In 2013, the facility was reopened under Vision Quest, a non-profit society for the treatment of addiction founded on the principle of abstinence. It serves adult men who are repeat offenders and uses the Twelve Step recovery program as well as Indigenous healing circles. It goes under the name of VQ The Creek.

When friends come to visit and they mention seeing a prison camp along Chilliwack Lake Road, I always make light of it.

"It's minimum security," I say, not 'fessing up to the serious nature of the crimes the inmates have committed.

I am reluctant to acknowledge that our retreat might be anything less than perfect. I don't want the prison world intruding into my life. I turn away from other social problems, too—homelessness, prostitution, runaways. They are harsh realities, I know. I donate to local charities and tell myself that I am helping. But emotionally, I put them at a distance.

Is my summer's quest nothing but a nostalgic tour of the past? How much unpleasant reality am I ready to face? By nature, I am a glass half-full kind of person. I focus on the positive. I prefer my world to be in harmony, not in discord, whatever the cost. My pursuit of the lake's history is making me question everything, even my own fancy pair of rose-coloured glasses.

It is ironic. I have spent the last ten years raising money for a school in Zambia. I visit every year and see first-hand the impoverished living conditions of the children that come to the school. I am ready and even eager to commit to helping people halfway around the world. But I haven't got personally involved here at home. While I was working as a lawyer and raising a family, I told myself that I didn't have time. When I had more time, I threw myself into raising money for a project in Africa. But now, this summer is forcing me to open my eyes to some of the harsh realities of lake history.

The painful truth is that Indigenous peoples constitute twenty-seven per cent of the prison population, but only three per cent of the adult population of Canada. Over a quarter of the prisoners in the prison camps established in the Chilliwack River valley during the 1960s were Indigenous. The ancestors of those prisoners were here long before my family. If they had their land, if they hadn't been torn from their culture, maybe they wouldn't be in jail. I fill out a volunteer application form for Reconciliation Canada. Maybe this is a place for me to contribute.

I've always thought of the lake as my refuge. My worries disappear the moment I look out on that serene funnel of water leading to the snow-capped mountains. I am carefree, in the truest sense of the word. Care free.

But is a refuge really possible today? Perhaps I feel carefree only because I push the harsh realities down deep where I can ignore them.

Is it necessary to bring the painful realities out in the open?

Maybe it is enough that my pulse slows, my breath steadies as I look out over the peaceful water. It restores me.

Floods, Storms and Earthquakes

Patrick and I have Fin, Roger and Will with us on a rainy weekend in late September. A curtain of water blurs the outline of the lake. The clouds hang heavy and dark. It is quiet on the water. Only a few fishermen putt along.

"Time to pull up the flowers in the barrels on the porch," I say to the boys. "If we all pitch in, we can get the job done in no time."

After grumbles and procrastinations, the boys don their slickers and gumboots. I watch them splash in the puddles on the path to the garage to get the wheelbarrows. Rainwater trickles down the back of my neck.

"The gutters need cleaning, too," Patrick adds, pointing to a cascade of water flowing off the roof at the edge of the porch.

"Maybe Kevin or Michael can get to it when the roof is dry."

Fin, Roger and Will yank out the withered plants and heap them in the wheelbarrows. A great job for the boys: no need for precision, just brute force. There's dirt everywhere, on their clothes and on the porch.

"Let me hose off the porch," says Patrick.

"Can you hose off the boys as well?" I ask.

We can hear the crescendo of Cupola Creek and see the outpouring at the mouth.

"The creek is a torrent," I say. "We should check the water system."

We put the dead plants on the compost, put the wheelbarrows away and head up the path along the waterline. The boys slip and slide on the path, racing each other to the top. Mud splatters across their faces.

"Why are we checking the water system?" Fin asks.

Roger calls out, "Fin, Will, come over here. I found some pieces of concrete."

We clamber over to where Roger is pointing at his find. He has discovered chunks of concrete hidden in the forest near the creek.

"That's part of our old water system. A long time ago, it was washed away during a rainstorm," Patrick says.

"That's why we're checking the water system," I add.

Our original water system was simple. Every year we dammed up a pool in the creek, put a filter on a plastic pipe and laid the pipe in the pool. Lengths of pipe on the ground carried water down through the forest a kilometre to the cabin. Water flowed out of the pipe into a washbasin. In time, the pipe was attached to taps at the kitchen sink. This worked for over forty years.

In 1965, the Ministry of Forests put a forestry road through our land. As part of the trade for a right of way, the ministry built us a new water system. Copper pipe all the way to the cabin, a modern filter, a concrete water tank on a concrete foundation. The new system was completed by the end of the summer.

"This water system will last a hundred years," Poppy said.

A heavy snowfall occurred the following winter from January through March. Unbeknownst to us, a snow and ice wall built up across the creek above our water system. In April, rains pummelled the mountains. The raging creek broke through the ice wall and obliterated everything in its path.

Mother, Dad and I arrived at the cabin at Easter to find large boulders and splintered tree limbs on the beach beside Cupola Creek.

"Look at all this debris," says Dad. "Something's happened up Cupola Creek."

"Let's take the water filter with us," I say, "so we can connect the system in the creek."

I go up the path with him. At seventeen, I know the ins and outs of our water system. But we're faced with devastation. We can't believe our eyes. Boulders and tree limbs are flung through the forest. We clamber over the debris and eventually reach the water system. Concrete and copper pipes are in a mangled heap alongside the creek bed. Deep channels have been gouged out of the hillside. The water system is gone.

"What a mess," says Dad.

"Oh, no. How are we going to tell Poppy?" I ask. "His new water system is wrecked."

"Every now and again Mother Nature likes to show us who's boss," Dad says.

"We can rebuild it, Dad. Or maybe just go back to the simple system we used before."

The old system functioned fine for forty years. As soon as we build a new system, it is destroyed by an act of nature. That summer we reinstate the same simple system we had before. It has lasted us for another fifty years.

We've added a few bells and whistles to our once-rustic summer life. We now have hydroelectric power and an internet connection. We have a washing machine and dryer. But we haven't tamed the wilderness. Every so often, just when we're getting smug, nature reminds us, reducing our bells and whistles to matchsticks in seconds.

Some natural disasters can't be fixed.

Every time we drive to the lake, Dad points out Mount Slesse, the highest peak in the Cascade Mountains we can see from Chilliwack Lake Road.

"Mount Slesse was the site of Canada's worst air crash," Dad says as we pass the site. The rest of us can mouth every word.

All I know is that one single fact. I don't know the story. I decide to find out what I can about the tragedy that occurred within thirty kilometres of the lake. I take out a book from the North Vancouver library, *Disaster on Mount Slesse: The Story of Western Canada's Worst Air Crash.*

On December 9, 1956, Trans-Canada Airlines Flight 810 is lost on a regularly scheduled flight from Vancouver to Calgary. This is early days in air travel. People are already nervous trusting in an aluminum tube to fly them from one part of the country to another. Having one disappear from the sky is earth-shattering. It will be five months before the plane is found and the fate of the sixty-two people aboard becomes known.

As the Canadair North Star sits on the Vancouver tarmac that December evening waiting for its passengers to board, the pilot is advised of severe weather predicted for the flight to Calgary. The cloud cover starts at 900 metres and extends to 6,000 metres. Extreme turbulence is reported around Hope. Winds over the mountains are clocked at a hundred and sixty kilometres per hour.

The plane is cleared for flight in the standard corridor. The fifty-nine passengers board the plane, and at 6:10 pm, it taxis down the runway and takes off. Immediately, the plane is rocked with turbulence. Forty-two minutes into the flight, the instrument panel flashes a red warning light. Something is wrong with one of the four engines. The captain requests clearance to return to Vancouver.

The last message from the captain received by air traffic control in Vancouver is at 7:10 pm. He establishes his location as having passed Hope. Radar contact with the plane is lost at 7:11 pm in the vicinity of Silvertip Mountain in the Cascade Mountains, southwest of Hope.

The aircraft should have landed at Vancouver by 7:40 pm. When the air traffic controllers cannot raise the pilots, they alert crews to the possibility of an emergency landing. At 8:00 pm, a search begins, the largest in Canada and the first ever high-mountain search for a passenger plane of that size. Rainstorms, gale force winds, sleet and snow continue through the night. Calls coming in to police headquarters in Chilliwack report a huge flash of light in the mountains near Cultus Lake during the early evening of

December 9. However, the search remains focused on Silvertip Mountain south of Hope.

The next morning, more than fifty members of the Chilliwack Flying Club turn out to assist in the search. At that time, planes do not carry emergency locators or "black boxes." A visual sighting of the wreckage is the only hope. But the storm is increasing in intensity. All small aircraft are grounded.

The first day passes with no sign of the downed plane. On the second day, more than sixty aircraft take to the air even though the weather is still stormy with little or no visibility. By the fifth day, the search is expanded to the ground with volunteers slogging through dense underbrush in the mountainous terrain around Silvertip Mountain. Days of heavy snowfall have laid a blanket of white many feet deep over the mountains. Even after all hope is lost, the search continues.

On December 23, a member of the Chilliwack Flying Club spots something glittering on the side of Mount Slesse. He thinks it might be a piece of fuselage. Again, this information is ignored.

Finally, on December 27, Trans-Canada Airlines issues a press release announcing that everyone on board Flight 810 has lost their life. The search is officially abandoned.

I stop at this point in the book, thinking what a painful Christmas this must have been for the families of the people on that plane. No evidence was found but they would know that survival in that landscape, under those extreme conditions, was impossible.

One of the searchers is Elfrida Pigou. Born in 1911 in the central part of the province, she hiked the hills of Vernon throughout her childhood. At a climbing school in Garibaldi, near Whistler, Elfrida experienced the magnetic pull of the mountains. She threw herself into this new vocation, discovering that, despite her diminutive size, she excelled at climbing. She was wiry, five feet and ninety pounds. Bookish, with wire-rimmed glasses, she was quiet and retiring, except when leading ascents of the tallest mountains in British Columbia. In the book, *Disaster on Mount Slesse: The Story of Western Canada's Worst Air Crash*, she described the allure of mountain climbing:

> On the mountains, there is no clash. The climbing itself is all rhythm and harmony, and wherever one looks there is also rhythm and harmony as though it had been designed by some very great artist.

Mount Slesse, "the Fang" in Halq'emeylem, was considered a sacred place by the Ts'elxwéyeqw. It was first climbed in 1927, following the southwest route, but the more technical ascent of the northeast buttress had never been

View of Mount Slesse in the Chilliwack River valley. In December 1956, Trans-Canada Airlines Flight 810 crashed into Mount Slesse, triggering the largest search in Canada's aviation history. All aboard were lost and the plane was not found until May the following year. Courtesy of the Chilliwack Museum and Archives, 2011.041.004.

achieved. Elfrida had set her sights on this goal. In the summer of 1956, Elfrida joined a group attempting the ascent of Mount Slesse's northeast buttress. Clouds set in and the attempt was unsuccessful. Elfrida vowed to return.

Elfrida follows the news of Flight 810's disappearance in the Cascade Mountains with interest. She's convinced that Flight 810 crashed on Mount Slesse. She plans a second ascent to search for the plane as soon as the snow recedes.

On May 11, 1957, three climbers, with Elfrida in the lead, arrive in the Chilliwack River valley at the base of Mount Slesse. Weather conditions are good. They start their ascent in dense forest, with loose rocks and melting snow in the lower alpine. At fifteen hundred metres, the ground is frozen, providing more secure footing. However, thick cloud moves in at higher elevations and heavy snow hampers their progress.

With poor visibility, Elfrida becomes disoriented. She turns up the wrong gully and only recognizes her error when she reaches a nearby ridge. Too late to retrace her steps, she forges on. Ahead of her, a large object stands out in the expanse of snow. As she comes closer, she sees that it's a piece of twisted aluminum close to a metre long bearing the initials TCA and a series of numbers. She has no doubt that this is part of Flight 810.

With daylight receding, the three climbers have to forego further exploration and head down the mountainside. Elfrida straps the hundred-and-twenty-pound piece of metal on her pack and descends the mountain. On the way out, she ponders what she should do, whom she should call.

Once in Vancouver, she phones Paddy Sherman, a fellow Alpine Club member who's also a reporter with the *Vancouver Province*. On May 13, 1957, five months after Flight 810 vanished, news breaks of its discovery, sparking a maelstrom of activity. Trans-Canada Airlines confirms that the piece of metal is part of the wing of Flight 810.

Two days later, a search expedition of climbers led by Elfrida divided into two groups and tackled both the east and west faces of Mount Slesse. Near the top on the west face, at 2,300 metres, they looked up and saw the nose of the plane dangling over the face of the peak, held in place by a tangle of control cables wound around an outcrop. The remainder of the plane had plunged 600 metres to a basin below.

One of the climbers reported, "If the plane had gone thirty feet to the right or about forty or fifty feet higher, the sixty-two people aboard could be alive today." How heartbreaking it must have been to hear such news.

It was not until late August that the snow conditions were stable enough for a climbing team to escort the coroner to the site. The impact of the crash had obliterated almost all recognizable traces of the bodies. After two days of searching, parts of seventeen bodies were recovered. A single grave was dug and the remains were buried together. With prayers, the coroner committed the bodies to the solace of the mountain.

Two days short of the first year anniversary of the crash, relatives gathered at the base of Mount Slesse to unveil a monument honouring the pilot, flight attendants, and fifty-nine passengers of Flight 810. A bronze plaque engraved with the names of the victims was mounted on the two-metre granite memorial.

Over the next thirty-eight years, the crash site was unprotected, open to vandalism and plundering. The relatives of the victims appealed repeatedly to the government to protect the site. After a campaign of letter writing, the Friends of Slesse group was invited to the British Columbia legislative assembly in Victoria to witness the approval of laws to protect the site. In May 1995, five hundred and eight-two hectares on Mount Slesse were created as a protected commemorative site, dedicated to the memory of the lives it had claimed. Although climbers and hikers are allowed, disturbances, such as mining, logging and recreational development, are prohibited. A sacred place, and now a protected place.

The Chilliwack River is prone to flooding in the fall. During the week of November 9, 1989, my Mother and Dad are closing up the cabin for the winter. They drain the water line, set traps for the mice, empty the refrigerator, and put antifreeze in the drains. They take one last look around, pack up the station wagon, lock the cabin door and head down the Chilliwack Lake Road. It has been raining hard all week. They are unsettled to see the river such a chaos of debris and waves.

"The river is higher than I've ever seen," says Mother.

They come around a bend. The road in front of them is gone. The river has eaten a swath of the roadway.

"We can't get through here," says Dad. "We'll have to go back to Centre Creek Camp. They'll know what's happened."

They drive into the prison camp and talk to the warden. He tells them that the road has been swept away in several places. The Tamihi Bridge across the river near Vedder Crossing has also been washed out. Helicopters have been ferrying shifts of guards to the prison camps for the past several days.

"We didn't think anyone was still at the lake," he says. "Isn't it late in the year to be at your cabin?"

"With such a mild fall," says Dad, "we just didn't have the heart to close up."

The warden phones down to the Royal Canadian Mounted Police in Chilliwack to relay the news that two old folks and a dog are stranded at Chilliwack Lake. Light is draining from the sky.

"It's too late for any rescue tonight. Do you want to stay here for the night?"

"No, we'll sleep at our cabin, thanks," says Dad.

"Okay, come back in the morning," the warden says. "A helicopter will take you out."

Mother and Dad get in the car and return to their cabin. They are back at the prison camp by nine. A helicopter appears and lets out four prison guards for their day shift. Mother climbs in and Dad lifts their Labrador retriever, Max, aboard and climbs in after him. They are in the air only a few minutes when the side door of the helicopter comes unlatched. Mother, seated next to the door, grips the door handle with one hand and Max's collar with the other. She calls out to the pilot. He looks back, nods his understanding and keeps flying. At an open patch of ground, he lands the helicopter and locks the door.

The streets of Chilliwack during the flood of May 1894. Boys provided a ferry service for those needing to cross the streets with their homemade rafts. Courtesy of the Chilliwack Museum and Archives, PO682.

"All locked up tight now."

Twenty minutes later, they arrive at the army base in Chilliwack.

My parents were lucky. If they'd been on the road when the washout happened, they could have been swept away by the river. Floods are a part of the history of the Chilliwack Valley.

Six years later, late in the fall of 1995, Patrick and I are closing up the cabin. We are standing on the porch taking a last look at the lake before driving back to the city. Suddenly, I notice the water receding from the shore.

"What's happening?" I turn to Patrick and we run down to the beach. We watch mesmerized as the water pulls back nine metres, unveiling the floor of the lake. Water sweeps across the dock first one way, then the other. The dock swings back and forth until the cable breaks that tethers the dock to the beach. It crashes up against the sand. The water surges forward and sideways creating a standing wave.

"Let's get out of here," says Patrick.

We call our Labrador retriever, Monty, and retreat to higher ground. We watch for about twenty minutes. Just as suddenly as they began, the waves die down. The water grows calm. The only remnant of the phenomenon we just witnessed is the dock reared up against the shore of the beach, its cables askew. We retrieve ropes from the shed and tie the dock as best we can to trees on the shore.

"Nothing more we can do," says Patrick.

"What was that?" I say, as we head back up to the cabin.

As soon as we get home, we check the news. An earthquake occurred in Washington State. We find a news report that water in swimming pools had sloshed back and forth and up over the edges.

"Well, we're right on the border, so maybe that's what happened," Patrick says.

"I would never have believed it if I hadn't seen it."

At the lake, we are subject to the capriciousness of the elements. We are not in a protected cocoon of civilization. Nature affects us. We learn lessons from these disasters and near-disasters and inconveniences. The world is never entirely in our control. Maybe never at all. It is discomfiting to realize that we are just humble humans. We think we own the lake but we don't. If anyone owns it, nature does. Poppy was right. We're just the stewards.

Expropriation

The letter is dated December 16, 1971 and is received two days before Christmas. It is from the Ministry of Recreation and Conservation, Parks Branch and addressed to Christopher Webb, Cupola Estates Ltd.

> We have recently been instructed by the Honourable W. K. Kiernan, Minister of Recreation and Conservation, to develop major park facilities at the north end of Chilliwack Lake. At present, a small forty-acre parcel of land is available to the public for camping and boat launching. This small site received over sixty thousand visitors in 1971.
>
> In view of the very high public demand for park facilities and lake access at Chilliwack Lake we will be incorporating an additional four hundred acres of Crown land within the present "park" area and are preparing development plans for two hundred campsites.
>
> Unfortunately, all but one hundred and twenty feet of the lake frontage at Chilliwack Lake is privately owned. The necessity to acquire these private lands to satisfy public recreational land requirements as well as to forego conflicting public use of private lands is self-evident.
>
> We would therefore be pleased to know if your property at Chilliwack Lake would be available for inclusion in this proposed provincial park. We would appreciate your response to this letter at your earliest convenience.

My grandfather frowns as he reads the letter and sits down heavily. His greatest fear has been realized; his land is under threat of expropriation. Poppy calls Mother and Dad and reads them the letter over the phone.

"Can they expropriate our land without our consent?" asks Poppy.

"I don't know," says Dad. "Let me talk to my lawyer."

Thoughts are whirling through Poppy's head. He should have bought the American Fishing Club when he had a chance. The government got a toehold at the lake and now they want more. The road was only put in nine years ago. If there were no road, this would not be happening. The government has all the power. He has none. It's David against Goliath. He's eighty-four. Does he have enough fight still in him for a battle?

Poppy sends his response immediately.

> The property is not for sale. It has been developed with two summer homes and several smaller buildings and is in full use by the principals involved.

Mr. Ahrens, director of the Parks Branch, replies in February 1972.

> I am disappointed with the news that the property was not for sale since there is very little useable public frontage on this portion of Chilliwack Lake. I am not in a position to conclude this matter and may wish to contact you further.

Through that winter and into spring and summer, the threat of expropriation hangs over the family. Dad and Poppy review the expropriation sections of the Park Act that Dad's lawyer delivers to his office. They read that the Minister of Recreation and Conservation could "for the purpose of the establishment or enlargement of a park" purchase or expropriate any land by payment of "reasonable value." Dad's lawyer reviews the Parks Branch letter and advises, "There is no provision for discussion or compromise on any proposed expropriation except as to the question of compensation."

This advice sounds like a death knell. They have no grounds to fight this. The government has every right to expropriate their land. The only thing up for discussion is the price tag.

Poppy has several discussions with Earl MacLeod, who received a similar letter. They both come to the same conclusion: their hands are tied.

They focus on the question of compensation. Maybe if the value of the property is high, the government won't expropriate? Dad hires Penny & Keenleyside Appraisals to determine what the property at Chilliwack Lake is worth. He instructs the appraiser.

> We have approximately 58 acres, an access road, a gravity water system, two large summer homes, three smaller buildings, and many improvements. We also have standing timber on the property.

I find a copy of the appraisal in the correspondence files. It concludes that the property in 1971 has a market value of three hundred thousand dollars. Poppy and Dad are concerned that the appraised amount is not high enough to be a deterrent to expropriation. By the beginning of the summer, the Parks Branch still has not responded.

At the time of the letter, I am teaching in Zambia. Intermittent mail is our only communication with Canada. Phone calls are prohibitively expensive. I am four months pregnant and focused on the imminent changes in my life. I don't hear about the threat to our property until my return. I am glad

The author's grandparents on a bench in front of Cupola Lodge, circa 1963. Courtesy of Webb family.

I didn't know. I would have freaked out, especially being so far away and unable to do anything.

In August 1972, Patrick and I return from Zambia bringing with us our three-month old son, Kevin. The morning after we land at the Vancouver airport, we are on our way to the lake. I can't wait to see it again. I let out a sigh of contentment the first moment I gaze down the lake after two years away. It looks the same as when I left. I think to myself, "It is the one thing that I can count on never to change." I carry Kevin to the edge of the water and dip his toes in the water.

"Here it is, Kevin," I say. "It will be your favourite place on earth, too."

Gran and Poppy are waiting for Kevin and me to arrive. I walk over to Cupola Lodge. They sit waiting on the bench under the birch tree. Poppy, eighty-five; Gran, seventy-nine. They look old and frail. Poppy is quiet and gentle as I place the baby in his arms.

The next day, Poppy shows me the expropriation letter. Although he hasn't heard anything further, he is still concerned that the government will take away our land.

I can't imagine losing our mountain refuge. It was the place that I yearned for while I was in Africa. I want to pass it on to Kevin and our future children. I don't want to think about the weight of public good versus private interests. Instead, I focus on the reunion with my family and with my special place. I persuade myself that there is nothing to worry about. I don't want to even consider the possibility. I stick my head in the sand.

"I am sure everything will turn out okay, Poppy," I say.

I read the letters in Poppy's correspondence file. Why did the Parks Branch take the initiative to increase the size of the Chilliwack Lake Provincial Park in late 1971? The Parks Branch letter states that there were sixty thousand visitors to the park in 1971. I search the archives of the *Chilliwack Progress*.

In 1971, Cultus Lake Provincial Park had more than three hundred thousand day visits. Cultus Lake is the closest park to Chilliwack Lake, much closer to civilization, about eight kilometres from Vedder Crossing. Cultus Lake was one of the most widely used parks in British Columbia.

In 1971, an initiative was undertaken at the south end of Chilliwack Lake. As part of a training exercise, the British and Canadian armed forces engineers completed a one-kilometre road and two bridges, creating a new park called Sapper Park in honour of the military engineers.

In August 23, 1972, the deputy minister references the problem of overcrowding at Cultus Lake.

> The further development of recreation areas such as Chilliwack Lake is indicated. The Gift of Sapper Park to the province as Canadian Forces Base Chilliwack's 1971 Centennial effort provides an excellent starting point. There is no quick and easy answer to the problem. With more leisure time than ever before, people want to travel and enjoy beauty spots such as Cultus Lake. Demand for parks and campsites will continue, and only a realistic park development program can cope with the problem.

Poppy's alarm increases with the election in August 1972 of the New Democratic Party by a landslide. The New Democratic Party, espousing social democratic principles and supporting everyman over the privileged few, is determined to right the wrongs of previous governments. An election statement catches Poppy's attention: the New Democratic Party vows to increase the number of parks in British Columbia.

Poppy's anxiety mounts. He feels as if the sword of Damocles will fall at any minute and there is nothing he can do. Months go by. Then one day, Earl MacLeod phones Poppy. He has good news.

It is leaden grey when I look out the window of the cabin this end-of-September morning. There are no mountains, just impenetrable clouds. I can't tell where the water ends and the clouds begin. A cluster of Canada geese floats in suspension on the water, their white chinstraps on black heads a muted reflection breaking the expanse of grey. The doleful drizzle continues. The leaves on the cottonwoods droop dejectedly. The sand is sodden. The geese huddle dispiritedly before propelling themselves up into the clouds setting their V pattern towards the south.

I shake off my doldrums, put on my rain slicker and gumboots and step outside. Patrick is working on a report and happy to stay put. I can hear the honking of the geese, a chorus fading into the distance. I see that Chris's car is in the drive at Cupola Lodge. I'll go and see him on my way back. I walk down the beach to check on the MacLeods. Making sure everything is shipshape. We have a higher risk of break-ins in the fall, with fewer people about. Plastic bag in my pocket, I pick up garbage along the way. I smile to myself thinking about Will and his enthusiasm for keeping the beach clean.

Coming towards the MacLeod cabin, I see smoke from the kitchen fire. I climb up the stairs to the porch, calling out, "Hello, MacLeods."

John MacLeod is in the kitchen.

"Come on in, Shelley," he says. "I'm fixing a leak under the sink."

"Hi John, how are you?"

"I could do with a break," says John. "How about some tea?"

We sit and sip our tea. He tells me news of his family. A wedding. The birth of a grandchild. The talk turns to the lake. He is thinking he might have to re-dig the hole for the outhouse. Maybe he'll get his grandson to help. They have an indoor toilet now but the outhouse is still used with enthusiasm by some of his family.

"Just like ours," I say. "It's Michael's favourite place."

"I can remember when your grandfather put in your outhouse," says John. "Two holes and a window. Pretty spiffy at the time."

"Some old things are treasures to hold onto," I say.

"Speaking of treasures, I've got some old photos that I want to show you," John says. "Just wait here a minute."

When he returns, he has a photo album of the MacLeods' first years at the lake. Black and white pictures, yellowing at the edges, black triangles holding the photos in place, show the MacLeod cabin going up log by log. John is eight, standing beside his dad, Earl MacLeod. A faded photo of my mother and John, who is standing in a replica of our cedar box used to preserve bedding.

"It was such luck that my dad flew your grandfather in to check on the fire in 1938," says John. "Otherwise, we would never have found this place."

"John, I've been looking through my grandfather's files. Do you remember when the government wanted to expropriate our land?" I ask.

"I sure do," John says. "My dad was in a state when he got that letter."

There were more pieces to the puzzle. In 1972, John was in discussions with the Parks Branch for other reasons.

"What was that about?" I ask.

"Ten acres of land at the outlet of the Chilliwack River," John says.

"I didn't know about that land, John," I say. "How does it fit into the picture?"

In 1943, Earl MacLeod wrote to the Department of Lands, to ask whether he could buy property on the Chilliwack River. He was granted a

The author's mother leans against a cedar box built for storing bedding for the MacLeod family, circa 1939. John MacLeod is standing inside the box. Courtesy of MacLeod family.

lease of ten acres. After leasing the parcel for a few years, the Department of Lands agreed to sell it to him. When John and Jean were married, Earl gave the land to them as a wedding present. John's idea was to subdivide the land and create lots for a recreational development. In the mid-1960s, he put in a rough road, developed a subdivision plan, and analyzed options for a water system. By the late 1960s, the Post Creek community was being put forward by the Forestry Branch. The provincial government did not see the benefit of another recreational development near Chilliwack Lake. John couldn't get approval to subdivide. The American Fishing Club sold their land to the government in 1971. About the same time, John was considering whether he should keep going or abandon his idea.

John was ready to give up when Earl MacLeod received the expropriation letter. John decided to use his ten acres on the Chilliwack River as leverage. He called the Parks Branch and offered to sell his property on the river. In return, John wanted the Parks Branch to promise that they were not going to pursue his parents' property on Chilliwack Lake to expand the park.

"I can remember quite clearly a telephone call with Mr. Ahrens of Parks," says John.

"He assured me 'that it is not the practice of the government to expropriate private lands for parks and that it was not their intention to do so for the

Chilliwack Lake Provincial Park". I remember this because I wrote it down and I still have those notes."

"Did you tell my grandfather about this?" I ask. I can remember, the fall after I had returned from Zambia, Poppy being agitated about the threat of expropriation. He talked about nothing else whenever I was with him. Even though I was trying to keep a brave face, I was getting worried too. Why hadn't we heard anything? What would we do if our land were taken away from us?

"Oh yes, my dad phoned and told him. There was a long, collective sigh of relief, I can tell you!"

"So you saved our properties from being expropriated. You're the hero!"

"Oh, I'm not sure I can take the credit. But maybe I nudged them in the right direction. And thank goodness we haven't had to worry about this since," he says.

"Oh, we'd fight it tooth and nail!" I reply.

The rain is still pouring down as I walk back to my cabin. I pop in to see Chris at his cabin and tell him the story of John's land sale to the government. He's surprised by the story, too. As I walk up the steps to my cabin, I ponder this new chain of events revealed by John. I had a vague recollection that John had owned land at the river but had no idea that John's offer to sell might have stopped the government in its expropriation tracks. John's ten acres might have made the difference. Funny, I don't remember this moment of salvation. All I remember is that suddenly Poppy was no longer anxious. He told me the government was not expropriating our land. The albatross had been lifted from around his neck.

Even so, this threat remains in the back of my mind. Every time we make an improvement, I think to myself, good, this increases the value; makes it harder for the government to expropriate.

The Parks Branch continued to expand the Chilliwack Lake Provincial Park. In 1980, they set aside ninety hectares of old growth cedar forest as an ecological reserve at the south end of Chilliwack Lake. In 1997, a major expansion to the park encompassed all of Chilliwack Lake and the surrounding mountains, a total of nine thousand hectares.

Expropriation is no longer threatened, but with this last expansion, our property is now completely encircled by the park. I don't know whether to be alarmed at the size of the park or relieved that the lake is now protected. Surely a larger park will mean more people. Some will appreciate and look after the environment. Some will litter and throw beer cans into the water. Some will light fires on the shore, putting us at risk of a forest fire. Some will respect our privacy. Some will not. In my heart of hearts, I wish the park had never been created in the first place. I feel uneasy about the future. Will the park be expanded even further? Will the lake become overrun with noisy boats and uncouth campers? I hate the film of uncertainty that lies over the place I love.

I wonder what the law says now. Could the government expropriate our land more easily? I search the online Statutes of British Columbia. In 1996, British Columbia enacted the Expropriation Act. The government still has the right to expropriate private lands, but must now pay "market value" rather than "reasonable value." A property owner has the right to ask that the valuation be reviewed by the British Columbia Supreme Court rather than a tribunal. It sounds as though the rules are tighter, more to the benefit of the property owner, but I don't know whether this decreases our risk of expropriation. Even with the changes, the property owner's only real bargaining power is the price, not the fact of expropriation itself.

What would we do if we lost our land? We would want a summer retreat for sure. It's the place where our family gathers, a place of traditions and celebrations, a place where we unwind and reconnect with each other and ourselves. Where would we go? We could not find anywhere with such beauty and isolation within a two-hour drive from Vancouver. I would be devastated.

I would fight. It would be my Armageddon.

I think about the role of private ownership in protecting our environment. Will increasing the size of the park result in preservation of the wilderness or just the opposite? Some believe that government ownership is necessary to preserve the environment, that private citizens just can't be counted on to be responsible. I agree that wilderness is an irreplaceable resource. We must protect what we still have. But is public ownership essential to that protection? I'm not convinced. Private groups, such as Nature Conservancy Canada, acquire land, plan conservation activities and manage and restore properties. These are non-profit charities fuelled by volunteers, not government. In fact, if my family is anything to go by, private ownership can accomplish that goal just as well, if not better, than government. Our family stewards our 27.5 hectares of wilderness with care. We occupy only a miniscule portion of our land. The rest remains pristine. We pick up litter. We douse campfires. We petition against clear-cut logging on the periphery of the park. We preserve our own property and encourage others to do the same. Poppy raised me to think of myself as a steward, rather than the owner of our land. I drill the same message into my grandchildren. I defy anyone, any government, any individual, to do a better job.

Closing the Cabin

We gather at the lake, combining Thanksgiving with our close-up weekend. This is earlier than our usual end-of-October close-up but it's the only time we can be here together with the grandkids' busy schedules. As Patrick and I arrive, a patchwork of shadows plays on the water, the sky an untidy congregation of cumulus, stratus, and nimbus clouds with peek-holes of pale blue. The chill of the wind foreshadows winter. The leaves of the cottonwoods are a kaleidoscope of yellow, green, gold and brown, each one waiting for the burst of wind that will send it on its dance to the ground. One by one, they twist and twirl, a slow pirouette.

Kevin and Michael arrive after dinner on Friday. Children jump out carrying pumpkins for carving, a Thanksgiving tradition.

The next morning, Kevin and Michael cover the picnic table with newspapers and cut the tops off the pumpkins. The kids plunge their hands into the centres, scooping up seeds and stringy innards. Joanna collects the seeds and toasts them in the oven. The kids draw their pumpkin faces onto pieces of paper, then with a black felt pen transpose their images onto the pumpkins. With small knives in small hands, the parents admonish the kids to be careful. Instead of a face, Nora carves her initial. Will's has a crooked smile.

The children place the pumpkins on stools on the porch outside the window. At dinner, we look out on six orange faces illuminated by candles, standing out in the dark of the night.

Two Cowichan sweaters are folded and sitting on the games table.

"Did you have them fixed, Rani?" asks Fiona.

She, Nora and Bronwen love these two old sweaters. One was Gran's and the other my mother's. They were unravelling at the cuffs. I had decided a month ago to see if I could get them restitched. My inquiry at the Coqualeetza Centre led to a knitter in Sardis. She said that the creamy grey colour of the cuffs was no longer available but she would try to find some old wool that was a decent match.

"They look brand new," says Bronwen, holding up the sweater for inspection.

"Look, she even put a wool patch over the hole in the sleeve."

"We have to keep these sweaters always," says Nora, the defender of traditions.

After pumpkin carving, the boys head outside for close-up jobs while the girls help me prepare the turkey. Fiona breaks the bread into small pieces for the stuffing. Nora cuts the celery. Bronwen chops the onions; after a few minutes she has tears in her eyes.

"My mum says that if you light a candle, your eyes won't sting," Fiona says.

She brings matches and a candle to the kitchen counter and lights the candle next to Bronwen.

"Don't think this is working," says Bron. "Nora, can you take over the onions?"

I melt butter in the cast iron pans, Nora adds the onions and celery and gives them a stir until they are limp. We put the stuffing ingredients together, adding salt and pepper and poultry seasoning last.

"How do you know how much to add?" asks Bronwen.

"I just go by instinct. Gran's rule was too little seasoning is always better than too much."

We get the turkey out of the fridge to bring it to room temperature. The girls are squeamish about handling the bird, except for Bronwen, who gamely gives it a wash and pats it dry with paper towels. At a pause in the preparations, the girls join the boys outside.

Fin and Roger, who slept in the boathouse all summer, bring the bedding into the cabin for storage and drape plastic sheets over the beds. They sweep the floor and bring in the battery lanterns, close the wooden shutters over the windows and padlock each one. The boathouse is finished, put to bed for the winter.

"I can't wait for next summer," says Roger. "This is the best bunkhouse."

"It's all ours, Rog," says Fin. "The girls don't want it."

Kevin and Patrick orchestrate the rest of the outdoor close-up jobs. The furniture is stored away under the porch. Michael has tied up the Valiant inflatable dinghy to the dock and is running the mercury five-horse motor to drain the gas. The hoses are drained and coiled and stored under the cabin. Will empties and cleans the plastic pails that we use to rinse the sand off our feet when we come up from the beach.

The weather turns. A downpour hammers the lake, wisps of clouds hide in the valley. The rain comes in sheets straight down, bouncing on the surface of the water.

The turkey goes into the oven, covered in tinfoil. It's time for our traditional Thanksgiving afternoon walk. With the rain still coming down with full force, I have a hard time generating enthusiasm.

"If we wait a few minutes, maybe it'll clear up," says Fiona.

No one is keen to head out into the rain, but we bundle up, nevertheless, donning gumboots, slickers and brimmed hats. The kids wear a motley assortment of headgear, a brimmed hat of Poppy's, a yellow slicker of Gran's. Dennis and Murphy run at full tilt, snapping at heels, shouldering each other into the bushes. Joanna looks like a drowned elf in a green army surplus poncho. Michael wears his bathing suit.

"No sense in getting my pants wet."

We spy an array of mushrooms, from lavender to pale pink to brown with white spots, showing off against a carpet of green. Water trickles down necks, into shoes, and onto the knees of jeans. We head through the campground, which also closes this weekend, to the river outlet, across the bridge over the Chilliwack River and up the Trans Canada Trail that leads to Mount Webb. We go partway along the trail and turn around to head back. Kevin and Michael talk about taking the kids on an overnight trip to Mount Webb, maybe next summer.

"We have other adventures for next summer, right, Rani?" says Fin.

"We still have a few trails to discover," I say.

"What about the wreckage of that plane that crashed?" asks Nora.

"On Mount Slesse?" asks Kevin.

"Yeah, that one. Let's go see the monument."

As soon as we open the cabin door, the aroma of roasting turkey assails our senses.

"Mmm, smells so good," says Nora.

"Who wants tea?" asks Kevin.

The kids and Kevin and Michael set up a game of *Settlers of Catan* on the games table, everyone with cups of tea. I sit in my corner chair looking at Gran's painting and listening to the sounds of the family bantering over the game's progress, cups clattering against saucers. I thumb through my notebook, reviewing the questions that I jotted down throughout the summer. For most of them, I have found the answers. There's always next summer for the rest. Murphy nudges my arm, bringing me out of my reverie.

"The game's done, Rani," Fiona says unusually loudly, and the kids look up at me, smiles of anticipation on their faces.

"Fiona, the turkey is done, the game is finished!" This grammatical error has long been a bugbear of mine. I think the kids now say it just to get a rise out of me. And I guess it works, because I correct them every time.

"Speaking of turkey, Patrick, can you take it out of the oven for me?" I ask, and he obliges.

"Can I baste it?" asks Nora.

"Sure," I say, passing her the baster.

"Let's leave the tinfoil off now so it can brown on top."

Molly is peeling the potatoes to be boiled, then mashed. Joanna is cutting up a turnip for her casserole, her own family tradition that she has brought to our Thanksgiving.

"I've actually learned to like turnip, at least Jo's turnip," Patrick says.

I bring out the cranberry sauce that I made the day before, substituting orange juice for water when cooking the cranberries.

We relax before dinner. Patrick has fallen asleep in the corner chair, a *Time* magazine splayed across his chest. I take apart the jigsaw puzzle and put the pieces back in their box. This one is of waterfalls; the one before it, a seaside view of one of the towns in the Cinque Terre. And the one before that, surfboards on a beach. Each one, a thousand pieces. Will helps me take the box downstairs where it will join myriad others sitting on the puzzle shelf, waiting for another summer season.

As I come back up the stairs with Will, it strikes me. One thing I learned this summer for sure. Our family settled here in 1922. That means we can plan our one hundredth anniversary.

"Hey, kids," I say. "Do you know that in eight years, we will have been here for a hundred years?"

"Let's have a party!" says Fiona.

"And invite everyone who has ever visited."

"All our friends."

"All the MacLeods."

"Maybe we should invite anyone who has every lived here," Patrick says, getting enthused about the idea. "What about descendants of Chief Sepass?"

"Maybe we could find a relative of that guy who drowned?" says Michael.

"Michael Brown. We could do a search in Ireland."

"Maybe we could put up a monument—to us!" says Will.

"Well, let's think about it. We've got time to get it organized," Patrick says.

"The turkey!" I exclaim, suddenly remembering that it should come out of the oven. "Patrick, can you take it out for me? It's *done!*"

Fiona sticks the wooden pieces in the shape of a turkey's head and tail in a large pumpkin and tops it off with a blue checked bandana around the head. The boys set the table, bringing in red and yellow leaves that Fiona adds to the pumpkin centrepiece.

Fin brings out the votive candles that we keep in the corner cabinet and spreads them around on the table.

"Can I light the candles, Rani?" Fin asks.

"Absolutely."

The table shimmers with flickers of light. Patrick sharpens the carving knife and carves the turkey, one platter for white meat, one for dark, spooning the dressing into a bowl on the side. Molly is mashing the potatoes, Joanna getting out the plates. I put my apron on to make the gravy.

"Can I help with the gravy?" says Bronwen.

"Sure. Here, put on this apron."

I drain off the drippings. Bronwen consults with me on the number of spoonfuls of flour to add. She stirs as I add cold water. I tell her the cold water is the trick to making gravy that is smooth, not lumpy. Next, we add the water from the potatoes and a broth made from simmering the neck and turkey innards. Gran's secret is Kitchen Bouquet. I add a few drops. It turns the gravy a rich brown colour.

"Okay, Bron, now taste it. Don't burn your tongue. Do you think the gravy needs salt and pepper?"

"I think so," says Bron.

"Okay, add some, but just a bit."

"I know, better too little than too much," she says.

I nod happily. Even the small traditions are passed on.

"Okay everyone, dinner's ready."

"It smells so good."

"I can't wait."

"Can I have a drumstick?"

We move around the central island in the kitchen, piling our plates high. When everyone is seated, Kevin says grace, giving thanks for family, present and absent, and the love and joy that surround us. He has always said grace in our family from the time he was a teenager.

"You know," Patrick says to the grandkids, "one of you will need to be able to say grace as you get older."

"Not me," says Fin. "I'll let Dad do it."

"You could always learn a simple one," I say. "Here's one I learned, 'For every cup and plateful, God make us truly grateful.' "

"Maybe I could learn that one," says Fin, not looking convinced.

After dinner, we sprawl on the couches in front of the living room fire. A game of charades ensues.

The next morning, Kevin makes breakfast with individual pancakes in the shape of each child's first initial. After breakfast, we start on the inside close-up jobs. The kids retrieve liquor boxes from the basement. We pack away anything valuable and store the boxes. Molly defrosts and cleans the fridge, placing the food in boxes and coolers to take back to the city. Joanna goes through the medicine cabinet, throwing away medicines that have expired and bringing to town any that can still be used.

Kevin and Michael take the boys up the water line to shut down the water system. Kevin goes into the creek and takes out the box filter and detaches it from the water line. Michael retrieves the line from the creek, and ties it up in the trees. He calls to Roger to turn the spigot that drains the water tank. The boys play in the woods waiting for the tank to empty. Will, the smallest, is lowered into the tank, towel in hand, to mop up the silt that has accumulated in the bottom of the tank.

"How are you doing, Will?" Michael asks. "Do you need another towel?"

"Yeah, there's a lot of gunk in here," he replies. "Can't Fin do this now?"

"No," says Fin. "I'm too big."

"How come when I want to do something, I'm not big enough? Now, 'cause I'm small, I have to do this yucky job."

"That's the way it goes. Keep up the good work, Scooch," Michael encourages, he and Kevin smiling.

Will finishes up the tank, they gather up the canvas bag, filter and muddy towels and traipse down the trail to the cabin.

Once the water is shut off, we are at the last stages of close-up. The taps are opened, the toilet tanks flushed and toweled out so no trace of water remains that could freeze and crack the ceramic. Antifreeze is added to the drains and toilet bowls. The boxes and coolers are piled on the back porch, the cars backed up and the gear packed in.

We stand in a line on the front porch.

"Good-bye, Lake," Fiona says. "See you next year."

"Good-bye, Lake," says Fin. "Roger, let's stay in the boathouse the whole summer next year."

"Yeah, and I want to finish our fort," says Roger.

"See you next year, Lake. For sure I'm going to be the first one in the water," says Bronwen.

"Bye, Lake," Will says. "Next year, I'm getting up on one ski."

"Can we do some more adventures next year, Rani?" asks Nora.

"We sure can," I say. I already have a list forming in my head.

Good-bye, Lake, I say silently. I'll be back.

The author's grandchildren's pumpkins carved for Thanksgiving, circa 2014. Courtesy of O'Callaghan family.

ACKNOWLEDGEMENTS

Throughout the writing of this book, I have had many occasions to witness the magnanimity of writers. Fellow writers who responded with generosity of spirit: Stella Harvey, Judy McFarlane, Anne Giardini, Michael Hetherington, Dan Francis, Megan Williams, Kathleen Pearson, and Leslie Hurtig. I feel privileged to consider myself part of this creative circle. I also benefitted greatly from the Simon Fraser University writing course on memoirs taught by Betsy Warland and the Sage Hill Writing Experience with mentors Merilyn Simonds and Wayne Grady.

Author Merilyn Simonds provided guidance and inspiration throughout my journey into the world of writing. She encouraged my hesitant first steps, cheered my progress, and pushed me to dig deeper, to be brave and to be bold. I owe her a debt of gratitude.

I appreciated the assistance of Shannon Bettles of the Chilliwack Museum and Archives and Tia Halsted of the Stó:lō Research and Resource Management Centre. There was nothing too arcane or obscure for their steadfast attention. Rachel Heide was relentless in her research into my grandfather's employment history at Library and Archives Canada in Ottawa.

Thank you to the MacLeod family, who share our ties of devotion to Chilliwack Lake. John MacLeod, who generously provided historical context to his family's history at the lake, and Flora Leisenring, who revealed a unique perspective in her stories about my grandparents. Marie and Dick Weeden provided insight into the pioneer Wells family of Sardis and gave me access to boxes of family photographs and mementos. Kris Sanders, whose father supported the early work of the Church Camp at Chilliwack Lake, gave generously of his time in responding to my inquiries about the camp's history and providing copies of early photos. Dan and Erin Coulter have been supportive throughout my research, providing links to Chilliwack Valley pioneers and the Post Creek Community.

Ann Mohs of Longhouse Publishing and Gerald Sepass, grandson of Chief William Sepass, opened my heart and mind to the complexities of local Indigenous history. *The Final Report of the Truth and Reconciliation Commission of Canada*, Volume One: *Summary, Honouring the Truth, Reconciling for the Future* was instrumental to my further education.

My publisher Vici Johnstone of Caitlin Press and production coordinator Demian Pettman have provided an equal measure of guidance and encouragement. Their attention to detail has resulted in a book of which I am proud.

Thank you to friends Jo Davidson, Mary MacKimmie and Susan Mawer, who were my touchstone; in particular, Susan, my first reader and cheerleader.

I started this project to learn about Chilliwack Lake so that my grandchildren would know the history of this place that means so much to me. This research prompted me to question my relationship with the environment, Indigenous peoples and the social dynamics of the family. And so my deepest gratitude goes to Patrick, my husband, and Kevin and Michael, my two sons, for their constant encouragement and unwavering love. In closing, this book is for my grandchildren, whom I entrust with the future of the lake. Thank you for giving me my quiet time for writing at the cabin when you really wanted me to come swimming.

Shelley O'Callaghan
North Vancouver, BC

Selected Bibliography

Anderson, Nancy Marguerite. *The Pathfinder: A. C. Anderson's Journeys in the West*. Victoria, BC: Heritage House Publishing, 2011.

Barman, Jean, Yvonne Hebert, Don McCaskill, eds. *Indian Education in Canada*, Volume 1: *The Legacy*. See chapter 6, "Separate and Unequal: Indian and White Girls at All Hallows School, 1884–1920." Vancouver: UBC Press, 1986.

Carlson, Keith Thor, ed.; Albert (Sonny) McHalsie, cultural advisor; Jan Perrier, graphic artist and illustrator. *A Stó:lō Coast Salish Historical Atlas*. Vancouver and Chilliwack, BC: Douglas & McIntyre and the Stó:lō Heritage Trust, 2004.

Carlson, Keith Thor. *The Power of Place, the Problem of Time: Aboriginal Identity and Historical Consciousness in the Cauldron of Colonialism*. Toronto: University of Toronto Press, 2010.

Chilliwack River Valley Historical Society. *In the Arms of the Mountains: A History of the Chilliwack River Valley*. Chilliwack, BC: Author, 2006.

Corley-Smith, Peter. *Bush Flying to Blind Flying: British Columbia's Aviation Pioneers, 1930–1940*. Winlaw, BC: Sono Nis Press, 1993.

Crosby, Rev. Thomas. *Among the An-ko-me-num*. Toronto: William Briggs, 1907.

Denman, Ron, ed. *The Chilliwack Story*. Chilliwack, BC: Chilliwack Museum and Archives, 2007.

Gilbert, Walter E., and Kathleen Shackleton. *Arctic Pilot: Life and Work on Northern Canadian Air Routes*. London: Thomas Nelson & Sons, 1940.

Halliday, W. M. *Potlatch and Totem and the Recollections of an Indian Agent*. London: J. M. Dent & Sons, 1935.

Harms, Kelly. "Charlie Lindeman: A Trapper and Prospector at Chilliwack Lake." Chilliwack, BC: Chilliwack Museum and Archives.

Jeffcott, Percival R. *Nooksack Tales and Trails*. Mount Vernon, BC: Sedro-Woolley Courier Times, 1949.

King, Thomas. *The Inconvenient Indian: A Curious Account of Native People in North America*. Toronto: Doubleday Canada, 2012.

Mackie, Richard Somerset. *Trading Beyond the Mountains: The British Fur Trade on the Pacific, 1793–1843*. Vancouver: UBC Press, 1997.

O'Keefe, Betty, and Ian MacDonald. *Disaster on Mount Slesse: The Story of Western Canada's Worst Air Crash*. Halfmoon Bay, BC: Caitlin Press, 2006.

Oliver, Jeff. *Landscapes and Social Transformations on the Northwest Coast: Colonial Encounters in the Fraser Valley*. Tucson, AZ: University of Arizona Press, 2010.

Sepass, William. *Sepass Tales: The Songs of Y-Ail-Mihth*. Recorded by Eloise Street. Chilliwack, BC: Sepass Trust, 1974.

———. *Sepass Poems: Ancient Songs of Y-Ail-Mihth*. Translated from the oral tradition by Chief Sepass and Sophia White Street. Mission, BC: Longhouse Publishing, 2009.

Stanley, George F. G., ed. *Mapping the Frontier: Charles Wilson's Diary of the Survey of the 49th Parallel, 1858–1862, While Secretary of the British Boundary Commission*. Toronto: Macmillan of Canada, 1970.

Wells, Oliver N. *The Chilliwacks and their Neighbors*. Vancouver: Talonbooks, 1987.

Wells, Oliver N. *Edenbank: The History of a Canadian Pioneer Farm*. Edited by Marie Weeden and Richard Weeden. Madeira Park, BC: Harbour Publishing, 2003.

Archives

Chilliwack Museum and Archives

Wells, Oliver. *Oliver Wells' Map of Indian Territory in 1858*. Edenbank Farm Fonds, 1966.

Wells, Oliver, interviewer. Tape 2008.033.004. Interview of Eloise Street on the Sepass poems. Marie Weeden Collection.

Chilliwack Museum and Historical Society—Oral History Subgroup. Interview of Carl Wilson. Additional Manuscripts 443, 1983.

Smith, Neil. *The Resurvey of the British Columbia and Washington State Border done in 1935.* Compiled from interviews with Gordon Watson.

BC Archives

Orchard Imbert, interviewer. Tape 720-1. Bob Joe interview, 1963.

Orchard Imbert, interviewer. Tape AAAB0824. Ray Wells interview, 1967.

Orchard Imbert, interviewer. Tape 2506 -1. "From the Mountains to the Sea: Patterns of the Valley," 1967.

Orchard Imbert, interviewer. Tape 728-1. W. B. Bailey interview, 1974.

Orchard Imbert, interviewer. Tape 3221-1. Walter Gilbert interview, 1978.

Stó:lō Library & Archives

Carlson, Keith Thor, ed. *You Are Asked to Witness: The Stó:lō in Canada's Pacific Coast History.* Chilliwack, BC: Stó:lō Heritage Trust, 1997.

Johnson, Ingrid. *History of the Chilliwack Tribe.* Field School Report. UBC Ethnographic Field School, 1995.

Knickerbocker, Madeline. " 'Bring Home the Canoe': History and Interpretation of Sepass Canoes in S'olh Temexw." Field School Report. Simon Fraser University, 2011.

Kostuchenko, Amber Dawn. " 'I am an Indian and live on the Indian Reserve': History, Culture, Politics, Colonialism and the (re)making of Chief Billie Hall." MA thesis, University of Saskatchewan, 2012.

Lerman, Norman, collector. *Lower Fraser Indian Folktales.* Compiled 1950–1951.

McKenna-McBride Royal Commission Meeting Minutes, 1913–1916; Chief Sepass's testimony, January 14, 1915.

Todd, Norman, recorder. *The Chilliwack Story of the SXWAYXWEY.* Told by Chief Richard Malloway.

Upper Stó:lō Interactions: Teachings from our Elders, 1983.

Library and Archives Canada

File 1898280. File 1898279. Christopher Webb employment records.

Department of Interior, Canada. Dominion Water Power and Reclamation Service. Water Resources Paper No. 51, Climatic Year 1924–25. Surface Water Supply of Canada, Pacific Drainage, British Columbia and Yukon Territory.

Department of Interior, Canada. Dominion Water Power and Reclamation Service. Water Resources Paper No. 106, Climatic Year 1946–48. Surface Water Supply of Canada, Pacific Drainage, British Columbia and Yukon Territory.

Diaries

Air Commodore Earl MacLeod flight log and diary. MacLeod Family Collection.

Ray Wells autobiography. Weeden Family Collection.

Government Agencies

BC Ministry of Environment, Lands and Parks, BC Parks Division. Management Plan July 2000, for Chilliwack Lake Park & Chilliwack River Ecological Reserve.

BC Ministry of Forests. Correspondence regarding proposed forestry road to Chilliwack Lake 1957, 1965–1968.

BC Parks Branch, Department of Recreation and Conservation. Correspondence regarding sale of land at Chilliwack Lake, 1971–1972.

BC Department of Lands. Correspondence regarding purchase of land at Chilliwack Lake by Frances Ellen Webb and Cupola Estates Ltd.

BC Agricultural Land Commission. Correspondence regarding application for release of land for prison proposal, 1994–1996.

BC Ministry of Forests. Correspondence regarding road to south end of Chilliwack Lake, 1965–1968.

Truth and Reconciliation Commission of Canada. *Final Report of the Truth and Reconciliation Commission of Canada.* Volume One. *Summary: Honouring the Truth, Reconciling for the Future.* Toronto: James Lorimer & Company, 2015.

INDEX

Page numbers for photographs are in boldface.

ABOUT THE AUTHOR

Shelley O'Callaghan worked as a volunteer teacher for two years with the Canadian University Services Overseas in Zambia, where she fostered literacy among adults in the rural communities and encouraged girls to stay in school and finish their education. She is passionate about history, social justice and the environment. O'Callaghan practised environmental law for twenty-five years and has been recognized as one of Canada's pre-eminent environmental lawyers. O'Callaghan is a member of the North Shore Writers Association, the Whistler Writing Society and the Canadian Creative Non-Fiction Collective. She attended the 2014 Summer Workshop of the Sage Hill Writing Experience and a writer's workshop with Merilyn Simonds in 2016.